For more information on this and other writing texts,
contact a Kendall/Hunt representative or visit these web sites:
www.kendallhunt.com
www.techreportguide.com

TECHNICAL REPORTS

A GUIDE TO STYLE, FORM AND DOCUMENTATION

R.A.O. Kleidon
Professor Emeritus
The University of Akron

KENDALL/HUNT PUBLISHING COMPANY
4050 Westmark Drive   Dubuque, Iowa 52002

Contributing editor: Anna Maria Barnum
Cover Design: Tim Klinger, Kleidon & Associates
Interior Design and Page Layout: Tim Klinger, Kleidon & Associates

Library of Congress Cataloging-in-Publication Data
Kleidon, R.A.O. (Rose Ann Osterman).
  Technical reports: a guide to style, form and documentation /
R.A.O. Kleidon.
  ISBN 0-7872-5060-0
  1. Authorship – Handbooks, manuals, etc.
  2. Technical reports – Authorship – Handbooks, manuals, etc.
I. Barnum, Anna Maria, contributing editor.   II. Title.
1998

Copyright ©1998 by Kendall/Hunt Publishing Company

ISBN 0-7872-5060-0

All rights reserved. No parts of this publication may be reproduced,
stored in a retrieval system, or transcribed, in any form or by any
means, electronic, mechanical, photocopying recording or otherwise,
without the prior written permission of the copyright owner.

Printed in the United States of America
10 9 8 7 6 5 4 3 2 1

CONTENTS

LIST OF ILLUSTRATIONS vii
FOREWORD ix
PREFACE xi

AN INTRODUCTION TO TECHNICAL REPORTS 3
    1.1 Formal and Informal Reports 3
    1.2 Reports at Work and in College 4
        1.2.1 A Difference in Purpose 4
        1.2.2 A Difference in Process 4
        1.2.3 A Difference in Sources of Information ... 4
        1.2.4 A Difference in Reward or Penalty 5
    1.3 Reports Categorized by Purpose 5
    1.4 Audiences of Technical Reports 6
    From the Editor: You, I or It? 7
    Plan Your Report: Thinking Ahead 7

INFORMAL REPORTS 9
    2.1 Memoranda 9
    2.2 Letters 11
        2.2.1 The Letter as Progress Report 11
        2.2.2 Letter Form 12

OUTLINING 13
    3.1 Outlining as a Tool for Thinking 13
    3.2 Traditional and Decimal Outline 14
    3.3 Outlining Principles 15
        3.3.1 Divide Logically 15
        3.3.2 Use Parts of Speech Consistently 15
        3.3.3 Favor Topic Outlines 15
    3.4 The Outline as Proposal 15
    From the Editor: Purpose, Scope and Audience 17
    Plan Your Report: Getting on Schedule 18

PARTS OF A REPORT: A BRIEF LIST 19
    4.1 Front Matter 19
    4.2 Body 19
    4.3 Back Matter 19
    4.4 How to Prepare and Assemble a Report 20

FORMAL REPORTS: FRONT MATTER 21
    5.1 Transmittal Correspondence 21
    From the Editor: Too Much of a Good Thing? 24
    5.2 Cover Labels 24

5.3 Title Page 24

5.4 Table of Contents 26

5.5 List of Illustrations 29

5.6 Abstracts 32

    5.6.1 Descriptive Abstracts 32

    5.6.2 Informative Abstracts 32

    From the Editor: Descriptive vs. Informative Abstracts 34

5.7 Glossary 35

5.8 List of Symbols 35

    Plan Your Report: Drawing a Map 37

FORMAL REPORTS: BODY 39

6.1 Introduction 39

6.2 Discussion 42

    6.2.1 Illustrations....................................... 42

    From the Editor: How to Identify a Table 42

    6.2.2 Subheads 44

    6.2.3 Sample Discussion Pages 45

6.3 Conclusions 52

6.4 Recommendations 52

FORMAL REPORTS: BACK MATTER 55

7.1 List of References.................................... 55

7.2 Appendixes 55

DOCUMENTATION 57

8.1 Quotations... 58

    8.1.1 Direct Quotations 58

    8.1.2 Paraphrases 61

    8.1.3 Plagiarism 62

    8.1.4 Wrenching from Context 63

8.2 Attributions 63

8.3 Parenthetic Citations 63

8.4 Reference Page Entries 64

    8.4.1 Typical Sources................................. 64

    8.4.2 Government Sources........................... 66

    8.4.3 Electronic Sources 67

    8.4.4 Unpublished Sources 67

APPENDIX: SAMPLE PAGES FROM STUDENT REPORTS 69

ILLUSTRATIONS

FIGURES

Figure 1 **A Proposal Report in Memo Form** 10

Figure 2 **A Progress Report in Letter Form** 11

Figure 3 **Spacing Guidelines for a Letter** 12

Figure 4 **Traditional and Decimal Outlines** 14

Figure 5 **Outline for *Warping*** 16

Figure 6 **Transmittal Memo** 22

Figure 7 **Transmittal Letter** 23

Figure 8 **Cover Label** 24

Figure 9 **Title Pages** 25

Figure 10 **Short Table of Contents** 26

Figure 11 **Long Table of Contents, First Page** 27

Figure 12 **Long Table of Contents, Continuing Page** 28

Figure 13 **Alternative Form for Listing Appendixes** 29

Figure 14 **List of Illustrations** 30

Figure 15 **List of Figures** 31

Figure 16 **Descriptive Abstract** 33

Figure 17 **Informative Abstract** 34

Figure 18 **Glossary** 36

Figure 19 **List of Symbols** 37

Figure 20 **Introduction** 41

Figure 21 ***Warping*, Discussion, Page 2** 45

Figure 22 ***Warping*, Discussion, Page 3** 46

Figure 23 *Warping*, Discussion, Page 4 47

Figure 24 *Warping*, Discussion, Page 5 48

Figure 25 *Warping*, Discussion, Page 6 49

Figure 26 *Marketing*, Discussion, Page 3 50

Figure 27 *Marketing*, Discussion, Page 4 51

Figure 28 *Marketing*, Conclusions 53

Figure 29 *Marketing*, Recommendations 54

Figure 30 *Marketing*, References 56

Figure 31 Excerpt from *Reinforced Reaction Injection Molding* 59

Figure 32 *Moving Freight*, Letter of Transmittal 70

Figure 33 *Digital Photography*, Table of Contents 71

Figure 34 *Moving Freight*, Abstract, Edited 72

Figure 35 *Digital Photography*, Discussion, Page 2, Edited 73

Figure 36 *Digital Photography*, Discussion, Page 3 74

Figure 37 *Finding an Empirical Formula*, Discussion, Page 2 75

Figure 38 *Finding an Empirical Formula*, Discussion, Page 4 76

Figure 39 *Synthesis and Analysis*, Discussion, Page 2 77

Figure 40 *Synthesis and Analysis*, Conclusion 78

TABLES

Table I   How Transmittal Correspondence,
Abstracts and Introductions Differ 40

Table II How Treatments of Tables and Figures Differ........................42

FOREWORD

Anna Maria Barnum, Contributing Editor

Technical Reports: A Guide to Style, Form and Documentation has been tested by hundreds of students at The University of Akron (most will know it as *A Stylesheet for Technical Reports*), where it has become a tradition and a valued reference book. Since the numbered reference system was first recommended here in 1976, it has become the favored system in many schools, businesses and publications. Students report that using the *Guide* provides a helpful transition to using a multitude of other style guides available.

PREFACE

Technical Reports: A Guide to Style, Form and Documentation helps the beginning technical writer by being

- simple, direct and easy to use;
- appropriate to a broad range of sciences and technologies;
- a sound basis for moving on to other forms.

Writers have not had style sheet guidance for very long. Even the venerable Modern Language's *MLA Manual* was first printed as recently as 1948. A good number of style guides have been written since, some by individuals, like Kate Turabian, and some by groups, like the Council of Biology Editors. Most are aimed at helping advanced students write theses or dissertations or helping professionals prepare journal articles. Many, like the *Associated Press Stylebook,* are written for a certain subject field.

The *Guide,* in contrast, is written for 2-year and 4-year students in science and technology programs as diverse as radiology, electronics and law enforcement. The *Guide* is also appropriate for use by graduate students. Although the recommended forms are academic in nature, the standards and process as well as most of the forms are applicable to professionals, too.

The *Guide* has several special features to help readers. Sample pages in the text and an appendix that collects examples of student work show how to handle quotations, illustrations, subheads and many other problems. A "✓" feature alerts readers to issues that should be discussed with an editor or professor. Blocks set in gray mark comments were written by editor Anna Maria Barnum in response to her students' questions during the nearly two decades in which she has used this *Guide* in class. "Plan Your Report" pages at the ends of Chapters 1, 3 and 5 help students define their projects, schedule their work and clarify their instructor's expectations.

Both formal and informal reports are shown here, though formal reports naturally take more pages. The *Guide* is captioned into quite small divisions to make the book easy to use as a reference. For the same reason, discussions are short and examples are many, with forty illustrations included. Since a guide to style should provide one clear, consistent format, this book does not show the range of forms possible or discuss the reasons a writer might choose one over the other.

- A writer on the job uses the form requested by the boss, company editor or client.
- An author submitting an article to a publisher uses the form preferred or required by the editor.

The forms shown here are recommended for three reasons: they are
- clear, simple and consistent;
- in widespread use today; and
- appropriate for use by college students, including beginning students.

Specifically, the *Guide* recommends a numbered reference documentation system with parenthetical citations, a decimal outline, and collegiate formats for the title page and transmittal correspondence. I have assumed the student will present a typed or computer-printed paper, so adaptations for commercial printing are not covered.

My thanks first to Dennis and Kurt, without whose patience, understanding and encouragement, there would be no book. Among my colleagues at the University of Akron, I owe a debt of gratitude to Anna Maria Barnum for editing, for years of classroom testing and for her belief in and support of this textbook; to Richard Fawcett, for a first draft of the parts descriptions and for many hours of valuable discussion; and to James Switzer for editing and encouragement. I appreciate also the permission from Linden Industries and McCann Plastics to use their expertise in polymers and from Wittenberg University student Betsey Wade to use her lab report as a student example. Lastly, my thanks to all my many students through the years, for whom this volume was written and who, in return, have taught me volumes.

R. A. O. K.

TECHNICAL REPORTS

A GUIDE TO STYLE, FORM AND DOCUMENTATION

An Introduction to Technical Reports

"To be able to be caught up in the world of thought – that is being educated." – Edith Hamilton

CHAPTER 1

As a guide to technical report style, this book devotes only brief passages to many general concerns that face writers, including technical writers.

You may wonder, for example, what technical language is. I will say only that this guide assumes technical language to be the language used in technical situations, such as reports, journal articles, papers, oral presentations, some promotional brochures, advertisements and textbooks. It even turns up in certain conversations. For more precise definitions, refer to a technical writing text or to your instructor.

This guide is limited to describing how to document technical research and how to present your findings in a written technical report.

> "There is no such thing as the 'right' form for a technical report."

1.1 FORMAL AND INFORMAL REPORTS.

By technical reports I mean certain forms in which much technical information is written. These forms can be formal or informal. Informal reports are called that because they lack the forms, or parts, such as title pages, tables of contents and others. Informal reports do not need these parts because they are short. Instead, they are written as memoranda or letters. Chapter 2 shows informal reports.

Longer reports need formal parts because the reader needs more help getting through them and because they are used differently by different readers.

While there is no such thing as the "right" form for a technical report, some forms are better than others. Good forms are inherently better because they are clear, simple, consistent and appropriate to the writer's purpose and audience.

Chapters 5, 6 and 7 show forms for formal reports. The *Guide* shows one form as a model, so discussion of forms is short and alternative forms are few. However, the form shown here is so simple and basic that it should be easy for you to adapt to whatever form you choose or (more likely) you are asked to use by your professor or editor.

1.2 REPORTS AT WORK AND IN COLLEGE.

Because technical reports written in college are practice for those written at work, the college student needs some sense of those work-related reports. Reports written at work differ from those done in college in several ways.

1.2.1 A DIFFERENCE IN PURPOSE. College reports are done to show the professor that you can report well and/or that you know the subject. *Work reports are written because someone needs to know the information.* The reader must build something, maintain something, buy something, or in some other way take action based on the information in the report. Thus, clear, unambiguous communication is more important than classic forms or linguistic subtleties.

Instead of trying to write long to meet a teacher's page requirement, writers at work try to accomplish their purpose in as few pages as possible. Brevity is always a kindness to the reader, and it creates a report that is cheaper to reproduce. Nevertheless, work reports are often longer than college reports, sometimes running to book length or even multi-volume series, simply because they cover so much more information.

1.2.2 A DIFFERENCE IN PROCESS. In school, your professor sometimes sees a report only when you turn it in for a grade. This means you have feedback only in comments accompanying a grade. In contrast, a report written at work must be revised until it does the job. Unclear instructions must be rewritten. Unexpected questions must be answered. Most large companies have editors to help you with these revisions. Sometimes you have access to a company art department for help with illustrations or in printing copies. Your goal of producing a functioning report must override your personal feelings of protection for your original version and of wanting to turn it in and be done with it.

1.2.3 A DIFFERENCE IN SOURCES OF INFORMATION. You may have field or lab work to report on in college, but, more often, you will rely heavily on other people's work. Some people will have reported their work in books or journals or web sites; others will have reported their findings in newspapers or magazines. What is not yet written and published you may be able to get by lecture, letter or interview.

Accessing this information means you must find it in

- your own texts (your personal library);
- university, public or company libraries;
- or by talking to the experts themselves.

A fairly extensive use of information taken from other people or their writing means using a formal system of documenting, capable of accurately identifying each source. Chapter 8 shows a documentation form for college use, one which can be the basis for using other documentation forms later.

Many technical reports done on the job use no library sources because they require information that is too recent or specific to be in the library. Reports that detail field experience or those that propose a business plan are examples of reports that use few, if any, library sources. Other writers at work use libraries regularly, including their own company libraries, to find source material.

In addition, any writer who submits an article to a technical journal wants to substantiate his work by citing the work of others, and that means using written sources. The editors of the journal he submits the article to will require him to use the documentation system they have chosen.

Whenever you have to cite sources for business purposes or for publication, the forms you learn in a university are good models, because university research uses the most detailed and reliable documenting. This means that no matter where you must next document your sources and no matter what variation in style may be requested, the principles of documented writing you learn here will still apply.

1.2.4 A DIFFERENCE IN REWARD OR PENALTY. A college writer will get a grade for his or her work. The writer can, to some extent, decide ahead of time the grade desired and thus the quality of work to do. On the job, poor quality work can endanger promotion, raises and even the position itself. After all, the company pays for the reports it requests from its employees, so it may fire those who do poor quality work and promote those who do high quality work.

1.3 REPORTS CATEGORIZED BY PURPOSE.

Because technical reports are written to satisfy the needs of a particular company at a particular time, there are thousands of variations. Categorizing by purpose, some of the most typical are proposals, feasibility studies, physical research reports, status reports and manuals. Manuals alone are a huge category, including operating manuals, calculation procedures, explanations of formulas or graphics, process descriptions, product functions, personnel procedures, among others.

Some of these reports reach a decision, then recommend action. Others do not reach a decision. This distinction is important because the design of the report may depend on whether the report writer must reach a decision. For example, a decision-making report needs conclusions and recommendations, while a report that, let's say, gives instructions, does not. On the other hand, a report without decisions may be meant for use as a reference book, and so benefits from having a long, index-like table of contents.

> **A PROPOSAL:**
>
> *A Proposal to Hawkin's Aluminum Corp. to Add Coil End Joiners to Roll Forming Lines 3 and 4 (Budget Included)*
>
> **A FEASIBILITY STUDY:**
>
> *Video-Conferencing for Physicians: Advantages and Disadvantages*
>
> **A PHYSICAL RESEARCH REPORT:**
>
> *Degeneration of Hue and Shade Perception in the Elderly*
>
> **A STATUS REPORT:**
>
> *Structural Integrity in Selected Ohio Bridges: Results of Field Testing, April-November, 1999*
>
> **A MANUAL:**
>
> *User's Manual for the AEX LaserJet 2 and 2M Printers*

Because of the difference in format for decision-making and non-decision-making reports, I show examples from a report in each category. The decision-making report used as an example is *Marketing via the Internet: A Preliminary Feasibility Study* and the non-decision-making report is a manual titled *Warping with a Sectional Beam.*

Marketing via the Internet is a feasibility study; as such its goal is to discover whether using the Internet for marketing is feasible (possible, practical, profitable).

However, it is also a preliminary study, so instead of reaching a decision on the question, it is limited to determining which issues need further study in order for the company to reach a decision.

Warping with a Sectional Beam is an instructions manual written to experienced, production handweavers to accompany new equipment that requires them to learn a new process.

1.4 AUDIENCES OF TECHNICAL REPORTS.

Lastly, a report is always heavily influenced by whether those reading it are, for example, company stockholders or factory workers. You must always consider your audience and write directly to them. Even a simple question like whether to define a word, for instance, can be answered only if you know the needs of your audience.

It is even more important in a classroom to imagine a real audience than in the working world, where the audience may be more apparent. As a college student you may never have given much thought to your audience, which is obviously unrealistic, not to mention impolite!

> "You may never have given much thought to your audience, which is obviously... impolite."

Writing to meet the needs of real people takes some adjustment, so it's important that you do some thinking about it and get some practice. *Real audiences need your information.* Otherwise, they wouldn't be reading your work. This makes them much more interested and, in ways, both more and less demanding.

In addition to having a large audience who reads to be informed, most writing has a second, separate audience: the editor. An editor can be anyone who reads your writing to check it, whether for accuracy, clarity, grammar or even spelling. The senior English major living down the hall could be your editor and so could your secretary. But the editor who really counts is the one who has the power to accept or reject your work: the teacher, the company editor, the editor of a journal.

Usually, though not always, the editor knows less than you do about your subject and more than you do about writing and presenting a report. An editor almost always wants you to do well, since the quality of your work reflects on him or her. To improve your work, an editor must pinpoint its faults. Though it is sometimes the editor's responsibility to correct those faults, more often it is yours. Since the writing must meet the editor's approval before it can be submitted to your real audience, you must be willing to work with an editor, even if that means making extensive revisions. (And even if it is painful!)

An Introduction to Technical Reports　　7

From the Editor: You, I or It?

You should feel welcome to use the writing in *Technical Reports: A Guide to Style, Form and Documentation* as an example on many points. (See the way capitalization and punctuation are used with bulleted points in this chapter, for instance.)

However, the way the *Guide* uses personal pronouns, while completely appropriate to this writing situation, is not likely to be the way you should use them. In the *Guide*, the pronouns *you* and *I* mean exactly what they say: *you* means you, the reader, and *I* means the author, Professor Kleidon.

In your own technical writing, you will probably wish to be less personal. Technical writing and really, all expository writing, is most often expressed in the third person (it, he, she and they) unless there is an exceptionally good reason to use the more personal forms. For example, in a brochure telling families of heart patients how best to cope with their loved-one's needs, it is both proper and advisable to use *you* throughout. Manuals intended for direct, hands-on use can properly be written from the "you" perspective. Most general and abstract manuals, as well as most other types of technical reports use, as a rule, the third person pronouns.

– AMB

PLAN YOUR REPORT: THINKING AHEAD

Answering these questions will help you define and plan your research and report. As you fill out this planning sheet, keep in mind that you should feel free to change anything as your project develops. At the same time, if you make too many changes, you will never finish the project. As you research and write, refer to your answers to keep your report on track and to alert yourself to any basic changes in your approach.

1. What is the subject of your report?

_____

_____

_____

2. Who is the audience for the report? ("My teacher" is the wrong answer! Re-read section 1.4.) Describe the audience as thoroughly as you can, being sure to estimate its reading level and its familiarity with the subject and its attitude toward the subject.

3. What does the audience need to be able to do as a result of reading the report?

4. Will the report be decision-making or non-decision-making? (Your answer to number 3 should tell you.) *Star your answer to this question.* It will be the basis for many more decisions.

5. Where can you find the information you need? Plan for a diversity of types of information.

Informal Reports

"The horror of that moment," the King went on, "I shall never forget it."
"You will, though," the Queen said, "if you don't make a memorandum of it."
– *Through the Looking Glass*, Lewis Carroll

CHAPTER 2

Informal reports are short reports written as memoranda or letters. In any business or industry, these are used for hundreds of reasons. Two of those reasons, to propose a project and to report on your progress, are especially applicable to a technical writing course. When you are assigned to write a long, formal report, you will probably also be asked to write a memo proposing the specifics of the report and one or more memos reporting on your progress. Here's how such a project works, in college and on the job:

When a formal report is requested, you will be told

- what information is needed,
- when it is needed (your deadline),
- how it is to be presented, and
- who is to receive your report.

Your first job is to do preliminary research to see what you can cover in the time allowed and to predict any difficulties in research, writing and presentation. You must check the literature in your subject for scope and availability, write a working outline and identify the report's purpose and audience.

> **QUICK CLUE:**
>
> Use Figure 1's *form* as a model for your memo, and let its *content* guide what you say in a proposal.

You then report these findings to your supervisor by proposing a specific report. When your writing plans have been approved, you proceed with research, writing and presentation. As you work, you write whatever progress reports your supervisor requires.

The proposal for your long report and the progress reports are informal reports. Such informal reports are written as memoranda or letters with detailed data, such as outlines, attached.

2.1 MEMORANDA.

A memorandum, or memo, is a special form of correspondence for use within an organization. While a letter goes from company to company, from individual to company, or in some other way communicates from one group or individual to another, the memo is made for using inside an organization.

Most companies give their employees a word processing file with these typical memo headings: company logo, department name and *to, from, subject* and *date* lines. Many companies also print memo forms with these headings for fast, handwritten notes. The memo example in Figure 1 shows a simple, typical format for you to use as a model. The content is a proposal for the report *Warping with a Sectional Beam*.

The subject line is neither a title nor a sentence. It is a noun or noun phrase that identifies the report as briefly as possible.

Paragraph 1 states the memo's purpose, then tells why the writer is qualified or interested in doing this report.

Paragraph 2 describes the planned report. Always mention attachments and enclosures.

Paragraph 3 tells whether enough sources are available.

Mention any special conditions.

Ask for the response you want.

Sign here or by the "From" line. Signatures can be more casual than a typed name.

Use "pc" for "photocopy." If someone is being sent a copy of your letter or memo, it is very important to let the addressee know this.

TO: Professor John W. Harris
      Associate Professor of English

FROM: Kurt A. Osterman
      Technical Report Writing 222:004

SUBJECT: Proposal for a formal report on sectional warping

DATE: February 8, 2000

For my formal technical report, I plan to give instructions for the process of sectional warping. Since I hope to be a production weaver after I graduate from the University's two-year textiles program, I have already learned how to warp a loom equipped with a sectional beam.

I am planning to report in four main sections: a general description, an explanation of the theory, a description of the mechanism, and instructions. An outline is attached.

At the library, I found three books and a journal article on parts of the subject. In addition, I own two textbooks that discuss sectional warping. My weaving teacher suggested I contact the Kent Weavers' Guild for the name of a local production weaver to interview. I also have the names of two companies that I will send letters of inquiry to. I have attached a list of sources.

As you said we could do, I have made plans to use this report in another course, Weaving for Production. I spoke to my instructor, Pamela Sharpe, who was agreeable to a joint project as long as I meet the requirements of both classes.

May I have your approval of these plans for my report?

Kurt

att: outline
   sources list
pc: P. Sharpe

FIGURE 1 A Proposal Report in Memo Form

The note "cc" for "carbon copy" is still in use even though carbon copies aren't. The newer "pc" for "photocopies" seems more appropriate now. Which form you use is infinitely less important than being sure you always tell the person you are writing to that someone else will also read the letter.

2.2 LETTERS.

Letters go to people outside an organization, so they need more information in the heading. They are also more formal. Most companies have printed letterheads, and they may also have rules about the signature they prefer you to use, the way to write the company name, and many other details. Some companies have extensive format and writing guidelines; others have none.

2.2.1 THE LETTER AS PROGRESS REPORT. Using memos in the classroom makes sense because students and teachers are working within one organization, but for practice, your instructor will probably want you to use a letter form for some class project. Figure 2 shows a standard business letter used as a progress report for *Warping*.

---

105 Magnolia Avenue
Carrollton, OH 44312

March 1, 2000

John W. Harris
Associate Professor of English
The University of Akron
Akron, OH 44325

Dear Professor Harris:

My work on *Warping with a Sectional Beam* is going well, with one change. As I began writing, I realized my outline was too complex. Everything planned for two sections (theory and description of the mechanism) can go in one section, as the attached revised outline shows. With an introduction and 10-12 pages of instructions, the report should still meet the 15-page minimum.

As we discussed, I wrote to two companies. Mailes Looms sent a brochure with two or three illustrations I can use. I have not heard from the Dick Blick Company. Also, I called the Kent Weavers' Guild president, who gave me the name of a local production weaver. I haven't scheduled an appointment yet. These sources, with the three books and two articles I have, should give me enough information.

Does the change in outline have your approval? Please let me know if you have any suggestions about how to proceed.

Sincerely yours,

Kurt A. Osterman

Kurt A. Osterman
TRW 222:004

att: outline
pc: P. Sharpe

Side annotations:

Unless you have letterhead, write your address here.

Use the date you send the letter.

Use a colon after the salutation, never a semi-colon.

Identify the report; mention the key point of the letter.

Be specific about changes or problems.

Mention attachments.

Report on the sources for your report.

Ask questions last. Request approval.

Identify the course and section.

No typist identification means the author typed the letter himself.

FIGURE 2 A Progress Report in Letter Form

12 *Informal Reports*

2.2.2 LETTER FORM. Positioning a letter on the page properly helps it to look professional. The spacing guidelines below will help you create a letter that looks balanced on the page. With a very short letter, achieving balance can be harder. If in doubt about where to position your letter, fold the paper in half. Slightly more of the body of the letter should appear on the upper half. Figure 3 gives you guidelines for spacing a letter. It also serves as another example of a progress report.

Allow 1½" for top and bottom margins and 1" for left and right margins.

Double-space here.

Double-space or more as needed.

105 Magnolia Avenue   *return address*
Carrollton, OH 44312

February 21, 2000   *date letter is sent*

John W. Harris   *inside address*
Associate Professor of English
The University of Akron
Akron, OH 44325

Double-space here.
Double-space here.

Dear Professor Harris: *salutation*

My work on *Warping with a Sectional Beam* is coming along well. I have completed my search for information and have six pages of rough draft written.

Single-space the body of the letter and double-space between paragraphs.

As you suggested, I wrote to Mailes Looms, who sent a brochure that will provide two or three illustrations. I also have four other sources, including three textbooks and a journal article.

Does my progress on *Warping* have your approval?

Double-space here.

Sincerely yours,   *complimentary close; cap the first word; end with a comma.*

3-4 spaces, depending on handwriting.

Kurt A. Osterman   *signature*

Kurt A. Osterman   *typed name*
TRW 222:004

Double-space or more as needed.

Use "att:" only if you have stapled or clipped something to the letter.

KAO:mj   *typist identification*
encl: 2   *enclosure notice*
pc: P. Sharpe   *distribution notice*

FIGURE 3 Spacing Guidelines for a Letter

Outlining

"Art and science cannot exist
but in minutely organized particulars."
– William Blake

CHAPTER 3

Even though an outline is not part of a formal report, nor of an informal report, it is a widely used planning device for both of them. You will probably be asked to attach an outline to the memo you write as a proposal of your report.

Almost always, an outline must be submitted to a supervisor for preliminary approval. After the report has been requested and its requirements explained, the writer plans the approach and scope of the report. He or she then presents these plans in a letter or memo accompanied by an outline. Sometimes the outline is submitted alone.

Often, a progress report is required while the technical report is being written. If so, the progress report is a letter or memo with an outline. The outline, with such revisions as occur during the actual writing, becomes the basis for the table of contents in the finished report. Even without a required proposal or progress report using an outline, you should see the relationship between this important part of the thought process and the final form of the report.

3.1 OUTLINING AS A TOOL FOR THINKING.

Some students have the strange idea that outlines are fancy, pointless formalities. Nothing could be further from the truth. Outlines exist in quite different forms, so it's not form that counts. They have been used for centuries by people in every field, from architecture to zoology, certainly not because they are fancy or pointless.

> **QUICK CLUE:**
>
> Use outlining as your second step in thinking. Use brainstorming, branching, etc., first to create and explore. Outlining first may stifle your thinking.

Why have outlines been so widely used? Whether you want to work with a big idea in your own head or to communicate it, "divide and conquer" is the way to rule. The larger and more complex an idea gets, the more likely you will need to partition it. By splitting it into smaller parts and handling one at a time you can keep yourself and your reader organized. A natural second step after dividing is to fill the main categories with the smaller parts belonging to each.

14 *Outlining*

Outlining is not important for itself but because it is the visible result of division and classification, a basic way of thinking. If you have been uncomfortable with outlining in the past, you may be using it too early or expecting it to do too much for you. Outlines reveal organization, and as such they an important tool for thinking, too important to do without.

3.2 TRADITIONAL AND DECIMAL OUTLINES.

If you are used to outlining in the traditional way, you may be surprised at the decimal outline (also called a multiple decimal outline). It is illustrated in Figure 4, which shows exactly the same information presented in two different outlines.

Decimal outlines can be expanded infinitely, which makes them appropriate for certain manuals, especially very technical ones. The same feature makes decimal outlines work well in situations where information keeps changing or growing. Legislation; criminal, zoning and safety codes; and personnel manuals are good examples. The decimal outline also has the advantage that every number, even one appearing in isolation some pages into a document, tells the reader exactly where he is within the structure of the document. The traditional outline is more familiar, and it may be less cluttered looking, but it requires learning the traditional scheme of Roman numerals, capital letters, and so forth.

Figure 4 compares traditional and decimal outlines that go down four levels. In a traditional outline, a fifth level would be headed with an Arabic numeral in paren-

---

UNDERSTANDING BOTANIC NAMES
Traditional Outline

I. Names indicate origins
  A. *Parrotia Persia* for a plant from Persia
  B. *Hydrangea* 'Anna' from Anna, Illinois

II. Names like *autumnal* indicate flowering times

III. Names commemorate people or events
  A. Tribute to a person
    1. *Primula forrestii* for discoverer Geo. Forrest
    2. *Rosea* 'Graham Thomas' for famous grower
    3. *Iris* 'Annabel Jane' as a personal honor
  B. *Rosea* 'Peace' a tribute to WWII's end

IV. Names indicate physical characteristics
  A. Color, like *purpea* for a purple-flowered plant
  B. Size or shape
    1. *Trifolia* for a three-part leaf
    2. For a tube-shaped flower: *tubiflora*
  C. Scents
    1. Usually indicates a perfume
      a. *Citriodora* for lemon-scented
      b. *Frutescens* for fruity-scented
    2. Nasty scented plants labeled *foetida*

Decimal Outline

1 Names indicate origins
  1.1 *Parrotia Persia* for a plant from Persia
  1.2 *Hydrangea* 'Anna' from Anna, Illinois

2 Names like *autumnal* indicate flowering times

3 Names commemorate people or events
  3.1 Tribute to a person
    3.1.1 *Primula forrestii* for discoverer Geo. Forrest
    3.1.2 *Rosea* 'Graham Thomas' for famous grower
    3.1.3 *Iris* 'Annabel Jane' as a personal honor
  3.2 *Rosea* 'Peace' a tribute to WWII's end

4 Names indicate physical characteristics
  4.1 Color, like *purpea* for a purple-flowered plant
  4.2 Size or shape
    4.2.1 *Trifolia* for a three-part leaf
    4.2.2 For a tube-shaped flower: *tubiflora*
  4.3 Scents
    4.3.1 Usually indicates a perfume
      4.3.1.1 *Citriodora* for lemon-scented
      4.3.1.2 *Frutescens* for fruity-scented
    4.3.2 Nasty scented plants labeled *foetida*

FIGURE 4 Traditional and Decimal Outlines

thesis and a sixth with a lower case letter in parenthesis. Are you unfamiliar with the concept of "levels"? It means levels of generality. For more information about outlining concepts, ask your instructor.

3.3 OUTLINING PRINCIPLES.

The principles of outlining apply, regardless of which form you use. Does it make sense to have outlining principles? Yes, because if you break one, you know your logic is faulty. The rules flag problem areas; they tell you what to double check.

3.3.1 DIVIDE LOGICALLY. This standard rule makes sense: If it has an "A" it must have a "B." It says that if you divide a category, your division must result in at least two parts. This is only logical; you can not divide anything without coming up with at least two parts. However, keep in mind that you are under no obligation to divide any category and that the divisions you create need not be similar or balanced. Outlines follow the *logic* of division and classification; that an unbalanced *appearance* may result is irrelevant.

> "Outlines follow the *logic* of division and classification; that an unbalanced *appearance* may result is irrelevant."

3.3.2 USE PARTS OF SPEECH CONSISTENTLY. The items in any one category should be grammatically consistent. If you start with nouns or noun phrases in one category, avoid changing to adjectives or verbs. This principle helps prevent illogical mixes.

3.3.3 FAVOR TOPIC OUTLINES. Most outlines are topic outlines (like the ones in this chapter). Some writers use full sentences, but that can be cumbersome. Occasionally you will see a word outline, where each entry is restricted to a single word, but that seems unnecessarily severe. When planning a report, topic outlines are especially practical, because phrases will adapt easily to use as subheads for the discussion part of the report. Topic outlines are also easier to write and understand. Use a topic outline unless you are specifically instructed to do otherwise.

One of the advantages of outlining is that outlined material can be enlarged or shrunk to fit the time or space you have. Once your material is organized, you can subdivide any level or add to any level. In the same way, you can delete detail without damaging the organization. If you want to be more specific, you can focus on any division and delete the rest.

3.4 THE OUTLINE AS PROPOSAL.

The outline shown in Figure 5 is for the report titled *Warping with a Sectional Beam*. It was prepared to be submitted along with the memo shown in Figure 1. Together, they are a proposal.

You could use a similar outline to propose your report. The extra elements shown above the outline itself may help you and your teacher know more precisely what you have in mind.

16 *Outlining*

✓ Ask your editor or professor whether a traditional or decimal outline is required. Or is it your choice?

Identify the report type here.

This information is not essential to an outline, but it helps clarify your writing plan.

The "Scope" corresponds with the outline.

Be as precise as you can about identifying your readers.

The report will have an introduction, but it is not part of the outline. Only the discussion section of the report is outlined.

Phrases used in the outline will become subheads in the report —unless the report changes before then.

Items in each division should be grammatically parallel, but this can change from one section to the next. 3.1, 3.2 & 3.3 are gerund phrases, while 3.1.1, 3.1.2 & 3.1.3 are noun phrases.

Warping with a Sectional Beam (a manual)

Purpose: The purpose of this technical manual is to explain how a sectional beam works, to show and explain the parts, and to help weavers adapt standard warp formulas to the more specialized process of sectional warping.

Scope: The report will describe sectional beaming, revised warp formulas and auxiliary equipment. A description of the mechanism will be followed by instructions. Illustrations will be included.

Audience: Sectional beams are used only by experienced production handweavers, who know weaving terminology and the warping process.

1 Theory of Operation

2 Description of the Mechanism
  1.1 The Sectional Beam
  1.2 The Spools and Rack
  1.3 The Tensioner

3 Instructions
  3.1 Adapting Standard Warping Formulas
      3.1.1 Plain Weave
      3.1.2 Twill Derivatives
          3.1.2.1 Overshot
          3.1.2.2 Opposites
          3.1.2.3 Crackle
      3.1.3 Unit Class Weaves
          3.1.3.1 Summer and Winter
          3.1.3.2 Atwater-Bronson

  3.2 Using Auxiliary Equipment
      3.2.1 Winding the Spools and Rack
      3.2.2 Threading the Tensioner

  3.3 Beaming
      3.3.1 Tying On
      3.3.2 Counting Threads
      3.3.3 Moving Ends into Threading Position
      3.3.4 Maintaining the Cross

FIGURE 5 Outline for *Warping*

Manuals also sometimes include parts, materials or equipment lists and/or sections on troubleshooting or safety.

From the Editor: Purpose, Scope and Audience

Various writers use outlines in different ways and at different times. A few even use outlines as a first step, to explore the subject the same way most writers use brainstorming or branching. The author of this book, Professor Kleidon, is one of those who uses an outline to explore:

"Outlining inhibits exploration for many people, but I like to explore and organize at once. To me, it's faster and more revealing. Perhaps one reason I can do this is because a feel for audience and purpose is second nature to me. Moreover, I allow my exploratory outlines to be undisciplined. The 'errors' that pop up are as revealing as the parts that feel right from the beginning."

Most writers find the outline restrictive when they are exploring a subject, so using it too early can be a mistake. Most writers use outlines to

- organize material into a writing plan,
- communicate that plan to an editor or instructor,
- guide the drafting (others draft freely),
- refine the organization of succeeding drafts, and
- serve as the basis of the table of contents and of subheads in the discussion part of the report.

Unless a writer uses an outline to explore, it is important to consider the report's purpose, scope and audience before creating your first writing plan outline. There is little point in dividing the discussion section of the report into its logical parts (which is all that outlining consists of) until the writer has considered the needs of the reader. How could the writer even figure out what should be in the report?

A fair amount of time should be spent considering the reader from several perspectives. Why did the reader want this report? What educational level have the majority of readers achieved? If the audience is a mixed one, it is only safe to use the least educated segment of the audience when deciding what to include and what level of language to use.

Students who have outlines due should consult the applicable sections in technical writing textbooks, especially those parts dealing with audience analysis. Also consult class notes. It may be helpful to discuss these considerations with other people. (It is easy to overestimate an audience's understanding of principles or jargon.) Then, after giving sufficient thought to the preparation, the report's purpose, scope and audience should be written out in prose form (complete sentences) before an outline is attempted.

Figure 5 shows how the outline may look after the writer has done a careful job of thinking through the preliminaries. Notice that good outlines do not show an introduction or conclusion as numbered parts. These parts will be added to the report after the discussion section has been written. However, the germs of the introduction do appear in the purpose, scope and audience statements.

– AMB

PLAN YOUR REPORT: GETTING ON SCHEDULE

Like all writers, technical writers start with brainstorming, branching and other exploration techniques. Typically their second step is to identify and focus on the "best" parts of their brainstormed lists – those that represent important ideas or parts of the subject. Step three is drafting and step four is editing. Steps three and four are repeated as needed until the writer is satisfied with the manuscript. The last step is polishing (a close edit for spelling and grammar) and presentation. Some writers doing research will use a survey of library sources as part of their exploration, but most will research between steps two and three. It is in step four that problems with your research are discovered, so be sure to identify your sources clearly so you can easily double-check your work and/or add to your source material.

To schedule your project, work backwards from your deadline. Be reasonable; allow your-self time to attend classes, do other course work, eat, sleep and enjoy weekends.

This is my deadline: _____

Allow 3-7 days after you finish writing for polishing and presentation. You will need to have your report copied and bound during this step.
I should finish writing by _____

Allow two-four drafts and at least a week for each drafting/editing cycle.
I should begin drafting by _____

Allow at least two weeks to find and read source material.
I should begin library research by _____

Give yourself at least a week and no more than two to explore the subject. Be sure you understand your assignment before you begin exploring. When you have define your subject, audience and purpose precisely enough to fill in Planning Guide 1, you have completed step one.
I should start exploring by _____

This schedule suggests you should start work on your project six-eight weeks before it is due.

Parts of a Report: A Brief List

CHAPTER 4

This list is very brief, so you can easily see the whole scope of a technical report. The parts are listed in the order in which they appear, and those that are used only as needed are marked "optional." Although a report can be organized many ways, this list shows one practical way. Check with your instructor about whether this represents the report as assigned in your class.

4.1 FRONT MATTER.

- transmittal correspondence (letter or memo)
- cover with a label
- fly leaf
- title page
- table of contents
- list of illustrations or more specific versions, like a list of tables and list of diagrams (optional)
- abstract or introductory summary
- glossary (optional)
- symbols page (optional)

4.2 BODY.

- introduction
- discussion
- conclusion(s) (optional in some reports)
- recommendation(s) (optional in some reports)

4.3 BACK MATTER.

- list of references
- appendix(es) (optional)
- fly leaf
- cover

> "Hold on to the transmittal correspondence, so it is not bound in with the report."

4.4 How to Prepare and Assemble a Report.

Choose a quiet, businesslike cover stock and a two-pocket folder in matching or coordinating colors. The report should be typed or computer-printed on good quality white paper with a smooth finish and reasonable opacity (no onionskin or vellum). Buy pressure-sensitive labels ($3\frac{1}{4}$" x $1\frac{1}{2}$" or larger). Type two identical labels while they are still on the backing.

Print two copies of your report, or have a second one made at a copy center. Have the report bound. Warning: hold on to the transmittal correspondence, so it is not bound in with the report. Add the fly leaves between copying and binding. Apply one label to the bound report and the other to the folder.

Put the report and transmittal correspondence into a folder, with the report in the right pocket and the transmittal correspondence in the left.

✔ **Check with your editor or professor about whether a different presentation format is preferred or required.**

Formal Reports: Front Matter

"The beginning is the
most important part of the work."
– Plato

CHAPTER 5

The examples shown in this chapter illustrate the two basic types of reports: the report of decisions and the informative report. Specifically, they are from a manual, *Warping with a Sectional Beam,* and a feasibility report, *Marketing via the Internet: A Preliminary Feasibility Report.* Most sample pages are shown only once, from either *Warping* or *Marketing.* However, when the part differs with the type of report, two examples are shown. Such parts include tables of contents and abstracts.

Everything before the introduction of a report is considered front matter. All front matter is numbered with a lower case Roman numeral at the bottom of the page, centered or flush right.

5.1 TRANSMITTAL CORRESPONDENCE.

To decide whether to use a letter or a memo for the transmittal correspondence, consider both the situation and the requirements of your course instructor or editor. In general, a memo is used for reports to be read and used inside your company or organization, and a letter is written when the report is submitted to someone outside your company or organization.

Here, a memo is shown first, then a letter. Whichever form you use, transmittal correspondence is not counted or numbered as a page of the report. Occasionally it is bound into the report, but this is considered a *faux pas* by most readers. Much more often it is simply enclosed in the same envelope or folder with the report (appearing first in either case), paper-clipped to the front cover or slipped inside the front cover.

✓ **Does your editor or professor require a memo or letter of transmittal?**

Compare Figures 6 and 7. The language is almost the same, but because the letters are written to different people, there are some differences. Notice also how a letter is set up when the writer is using stationery with a letterhead.

22 *Formal Reports: Front Matter*

The first paragraph tells the reader the purpose of the memo or letter. The second paragraph describes the report and gives the reader any additional explanation he may need or want. Use third and fourth paragraphs for extra information. These center paragraphs may include

- identification of the audience.
- mention of any special features of the report.
- explanation of why coverage that may be expected is not included.
- explanation of any change since the last communiqué with the reader.
- brief history or theory if this reader needs it.
- explanation of the organization of the report if that may raise questions.

Spell names and titles accurately.

Identify yourself by course name, course number and section number.

Use the date the report is submitted.

The body is single spaced; paragraphs are not indented.

A very short letter or memo may be double-spaced so it looks better on the page. If you double-space, indent the paragraphs five spaces.

Side margins should be 1"-1½". Top and bottom margins should be at least 1½".

The initials "KAO" stand for the writer; the lower case initials "mt" stand for the typist. Do not include this line if you typed the letter or memo yourself.

"Encl:" alerts the reader that there is an enclosure. "Warping" is italicized because it is a short form of the report title.

TO: John W. Harris
           Associate Professor, English

FROM: Kurt A. Osterman
           Technical Report Writing 222:004

SUBJECT: Report titled *Warping with a Sectional Beam*

DATE: April 6, 2000

Please read the enclosed manual about the process of warping with a sectional beam.

This report describes the equipment, reviews the theory of operation and gives instructions for calculating a warp and putting it on the beam. The report is intended to be clear to experienced handweavers using any sectional beam. Because the reader may be reviewing the process as a step towards deciding whether to buy the equipment, an appendix listing sources for looms with sectional beams is included.

I thank Charles Glover, owner of Glover Supply, Inc., the U.S. distributor for Glimakra Looms, and Judy Parrish, who owns a LeClerc loom with a sectional beam, for letting me interview them and photograph their equipment.

Kurt

KAO:mt
encl: *Warping*

FIGURE 6 Transmittal Memo

- mention of the conditions under which the report was written.
- mention of previous correspondence with the reader.
- mention of sources that were particularly helpful.
- thanks to anyone who gave you special help, except the teacher or typist.

If approval must given or some other action taken, ask for it in the last paragraph. (Asking a teacher to grade a report is unnecessary.)

If the report was commissioned by another company and is being delivered to them, a letter format should be used. An example of a letter of transmittal follows. This example is done in full block style (flush left throughout) because the placement of the letterhead demanded it, but either a full block or modified block style is appropriate.

The Wordsmith Workshop

RESEARCH • WRITING • EDITING
123 FIRST STREET, AKRON, OH 44000

April 4, 2000

Ms. Barbara Getty, President
Kent Weavers' Guild
6321 Stone Road
Medina, OH 44256

Dear Ms. Getty:

Please read the enclosed manual, *Warping with a Sectional Beam,* commissioned by the Guild on November 19, 1999.

This report describes the equipment, reviews the process and gives instructions for calculating a warp and putting it on the beam. The report is intended for experienced handweavers using any sectional beam. Because the reader may be reviewing the process before a purchase, a list of manufacturers is appendixed.

I thank Charles Glover, the U.S. distributor for Glimakra Looms, and Judy Parrish, who owns a LeClerc loom with a sectional beam, for letting me interview them and photograph their equipment.

Please let me know if you approve this report or would like changes to be made.

Sincerely,

Kurt A. Osterman

Kurt A. Osterman

KAO:mt
encl: *Warping*

This is the date the report is submitted.

Put a short title right after the name. Put a long title on the next line. Compare with Figure 6.

Note the change to compensate for the lack of a subject line.

Mentioning the contract date is typical in business, rarely seen in schools.

Remind the reader of the next step and ask for action.

Allow 3-4 lines, depending on the size of your signature.

FIGURE 7   Transmittal Letter

From the Editor: Too Much of a Good Thing?

Compare the wording in Figures 6 and 7. In many instances they are identical, but there are some differences that relate to their different purposes. Find and study these carefully.

There is a danger in following these forms too closely. Notice the language starting these examples, and be aware that it would be safer not to imitate that language but to think up your own way to begin the transmittal. Otherwise the region's employers are going to become very tired of looking at transmittal correspondence that begins, "Please read...."

– AMB

5.2 COVER LABELS.

The cover label for *Marketing via the Internet: A Preliminary Feasibility Study* shows how to handle a two-part title. Make two identical labels and use one on the report and one on the folder. Buy labels large enough to allow plenty of room for all the information you need. If you must shorten, the subtitle or pre-title are the only parts you are allowed to omit.

Marketing via the Internet
A Preliminary Feasibility Study

by
Marise Jordan Strong

TRW 222:002   March 18, 1999

If you can distinguish between the title and subtitle by a line break and a difference in type handling, omit the colon between them.

Subtitles and pre-titles can be dropped from the label if you run out of room.

Show the course and section number.

This should be the date on which the report is submitted.

FIGURE 8   Cover Label

5.3 TITLE PAGE.

The title page is actually page i, and it is counted, but the page number is not shown. A good title page is visually pleasing. Items on it should be centered horizontally and balanced vertically. Avoid fussy type; use conservative type handling like the examples shown in Figure 9. Notice also that Figure 9 itself shows how to label multiple images as parts of one figure.

Three title pages here are shown in collegiate format. On the job, the person in your own company (and his or her title) or the name of the other company that commissioned the report is given prominence and the writer's name is omitted, as in Figure 9.4.

Formal Reports: Front Matter **25**

Marketing via the Internet
A Preliminary Feasibility Study

by
Marise Jordan Strong
Technical Report Writing 222:002

for
John W. Harris
Associate Professor of English
The University of Akron

March 18, 1999

9.1 Times in 30, 20 & 14 point

Warping with a Sectional Beam

by
Kurt A. Osterman
Technical Report Writing 222:004

for
John W. Harris
Associate Professor of English
The University of Akron

April 6, 2000

9.2 Helvetica in 28 and Times in 14 point

```
        MARKETING VIA THE INTERNET
         A Preliminary Feasibility Study



                    by
          Marise Jordan Strong
     Technical Report Writing 222:002


                   for
            John W. Harris
     Associate Professor of English
         The University of Akron




             March 18, 1999
```

9.3 Courier in 12 or 14 point

Warping with a Sectional Beam

for
The Kent Weavers' Guild
April 4, 2000

The Wordsmith Workshop

RESEARCH • WRITING • EDITING
123 FIRST STREET, AKRON, OH 44000

9.4 A Professional Title Page with Logo

FIGURE 9   Title Pages

26 *Formal Reports: Front Matter*

5.4 TABLE OF CONTENTS.

The two kinds of tables of contents pages being used in reports reflect two theories. Some people believe the table of contents (TOC) should list only major headings and should be only one page long, so that the whole scope of the report and its major headings are visible at a glance. Figure 10 is an example of a one-page table of contents. Others think a table of contents should show every (or almost every) caption used in the body of the report, to make small sections easy to find and to outline the entire report. Figures 11 and 12 show a long TOC.

Technical writing texts show this difference in their own contents pages. For example, Philbin and Presley limit the table of contents in their 547-page book, *Technical Writing: Method, Application and Management,* to two pages. On the other hand, William S. Pfeiffer lists every heading used in his 616-page text, *Technical Writing: A Practical Approach.* The result is a six-page table of contents.

Use this short form for decision-making reports and for reports with indexes.

Change the page number from lower case Roman numerals to Arabic numerals where the body of the report begins.

Write the headings on this page and in the report in exactly the same way.

The first page number to appear in the report is ii.

CONTENTS

```
LIST OF TABLES......................................iii

INTRODUCTORY SUMMARY...............................iv

INTRODUCTION........................................1

DISCUSSION..........................................2

I. MARKET REACH ....................................2
   A. Cross-Referencing to Internet Data ..........3
   B. Regional Markets on the Internet ............4

II. RETURN ON INVESTMENT ...........................6
    A. Costs of Web Sites and Servicing ...........9
    B. Costs of Support Advertising ..............11
    C. Fulfillment Costs ..........................12
    D. Effect on Pricing ..........................14

CONCLUSIONS........................................16

RECOMMENDATIONS....................................17

REFERENCES.........................................18
```

ii

FIGURE 10 Short Table of Contents

You can see that the size of the report does not tell you which type of table of contents to use. Some writers use both, as Killingsworth does in *Information in Action,* which has a one-page "Brief Contents" followed by a complete "Contents" listing that takes eight pages.

I recommend you use a one-page table of contents except in manuals. When a report is designed to be read straight through, or at least by sections, it needs only a one-page table of contents, but, because manuals are often used piecemeal, a complete table of contents makes sense with them. Even manuals do not need long table of contents when an index is included, an option that some word processing programs now make reasonably easy.

✓ Ask your editor or professor which table of contents is suggested or required.

CONTENTS

LIST OF ILLUSTRATIONS iii

ABSTRACT iv

INTRODUCTION 1

DISCUSSION 2

1 THEORY OF OPERATION 2
   1.1 Adapting Standard Warp Formulas 3
      1.1.1 Calculating Total Yardage 3
      1.1.2 Calculating Yardage per Spool 3
   1.2 Using the Mechanism 4
      1.2.1 The Sectional Beam 6
      1.2.2 The Tensioner 6
      1.2.3 The Spool Rack 7

2 INSTRUCTIONS 8
   2.1 Using Sectional Warp Formulas 8
      2.1.1 Plain Weave 10
      2.1.2 Twills 10
      2.1.3 Twill Derivatives...10
         2.1.3.1 Overshot 10
         2.1.3.2 Opposites 11
         2.1.3.3 Crackle 11
      2.1.4 Unit Class Weaves 11
         2.1.4.1 Summer and Winter 12
         2.1.4.2 Atwater-Bronson 12

ii

Use this long form for non-decision-making reports that do not have an index.

Indenting the page numbers from the right for fourth and fifth levels makes the page neater and easier to read.

FIGURE 11 Long Table of Contents, First Page

28 *Formal Reports: Front Matter*

Figures 11 and 12 show a long table of contents for the manual, *Warping with a Sectional Beam*. Compare the long and short tables of contents closely, and notice that besides length, there is another important difference. The feasibility report table of contents in Figure 10 has both a Conclusions section and a Recommendations section, while the manual TOC in Figures 11 and 12 has neither. Generally speaking, a decision-making report has conclusions and recommendations sections, but a non-decision-making report has neither. Occasionally, an instructor or editor will require a conclusion(s) section on a non-decision-making report. Check with your instructor or editor about this.

Figure 12 shows how to set up continuing pages of contents listing.

Notice that no conclusion is listed on this table of contents. That is because, as a manual, the report has no conclusion.

2.2 Using the Equipment 13
    2.2.1 Winding the Spools 13
    2.2.2 Loading the Rack 14
    2.2.3 Maintaining the Cross 15

REFERENCES 17

APPENDIX 18

iii

FIGURE 12 Long Table of Contents, Continuing Page

If your report has more than one appendix, show the last section of the table of contents like this:

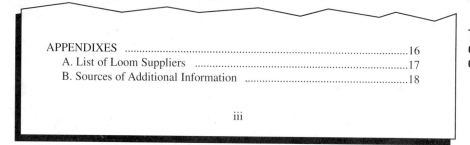

FIGURE 13 Alternative Form for Listing Appendixes

Create your table of contents after you have written your report, and base it on the headings actually used in the report. Be sure that the table of contents uses exactly the same wording as that which appears in the report itself. If you reduce the headings to fit a single page, omit a whole level at a time. Omit as many levels as you need to, so long as you do not leave out a major heading.

5.5 LIST OF ILLUSTRATIONS.

The idea of a list of illustrations is simple: it is a kind of table of contents to guide the reader to a graphic rather than a written element of the report. A list of illustrations shows the number and title of each illustration and the number of the page on which it appears. The more illustrations a report has, the more necessary a list of illustrations page is.

The heading should be as specific as possible. For example, if a report has diagrams but no other type of graphic, the page would be titled "DIAGRAMS."

If a report has several types of illustrations, they can be listed together under a more inclusive heading. The two such possible headings are "ILLUSTRATIONS" and "FIGURES." The term *figures* includes all graphic elements except tables. Graphs, diagrams, photographs, maps, drawings and charts are figures. A report with, for example, two graphs, a photo and a diagram would have a page titled "FIGURES."

The word *illustrations* is even more inclusive, since it means all graphic elements, including figures and tables. Since *illustrations* means both figures and tables, a report with two drawings, a diagram, a photo and three tables would have a page with the heading "ILLUSTRATIONS." Since a page headed "ILLUSTRATIONS"

✓ **Does your professor require illustrations? If so, how many? How many illustrations may a report have before your professor requires a list of illustrations?**

30 *Formal Reports: Front Matter*

Traditionally, figures are given Arabic numerals, and tables are given Roman numerals.

Tables are always listed separately and second, either below a subheading, or, if there is not enough space on one page, on their own page.

ILLUSTRATIONS

FIGURES

1 Sectional Beaming with Warping in Progress 5
2 A Sectional Beam 6
3 A Tensioner 7
4 A Horizontal Spool Rack 7
5 Threading the Tensioner 14
6 Beaming 15
7 Moving the Ends 16
8 The Glimakra Sectional Beam 19

TABLES

I Yarn Yardage per Pound 8
II Warp Yardage 9
III Real and Adjusted Warp Widths 9

iv

FIGURE 14 List of Illustrations

always lists at least two types of graphic elements, it has subheadings. If a report has tables and just one kind of figures, let's say maps, the maps would be listed first under the subheading "MAPS" and the tables would be listed second under the heading "TABLES." Subheadings are written in all capitals and are placed flush left. An example of a list of illustrations page is shown in Figure 14.

Formulas and lists are not considered illustrations. They are not given captions or numbers where they appear in the body of the report, nor are they listed on a list of illustrations page of any type. An exception is that in a report with a great many formulas, they may be numbered.

If you are separating the list of illustrations because there are too many to put on one page, do it according to category. For example, put a list of charts on one page and a list of tables on the next. You can use any of these three categories:

FIGURES

1 Sectional Beaming with Warping in Progress 5
2 A Sectional Beam 6
3 A Tensioner 7
4 A Horizontal Spool Rack 7
5 Threading the Tensioner 14
6 Beaming 15
7 Moving the Ends 16
8 The Glimakra Sectional Beam 19

iv

> **This list includes a chart, drawings and photographs. If you have too many to fit on one page, divide by categories and list one category per page with a specific heading on each page.**

FIGURE 15 List of Figures

figures, any specific kind of figure, or tables. These pages have no subheadings. Figure 15 shows a list of figures.

Just as Figure 15 shows a number, title and page number for each figure, so must the figure as it appears in the report have that same number and title. Of course, it must also appear on the page listed. Obvious as this is, it is an easy mistake to make. As you revise the report, you will move illustrations to keep them close to the text discussing them. Be sure to proof the page numbers on the list of illustrations as you finish your work on the report. To remind yourself to do this, you may want to pencil in page numbers or leave them off entirely until the body of the report is finished.

Illustrations can be one the trickiest parts of a report to handle smoothly. This part of the *Guide* discusses only the list of illustrations page and its variations. Refer also to section 6.2.1 for tips on how to create report pages with illustrations.

5.6 ABSTRACTS.

The abstract is the most carefully written and most frequently read page of a report. It is the shortest possible complete description of the contents and sometimes of the information in the report. The two contradictory goals of being brief and being complete make an abstract hard to write but very useful to readers who can tell quickly whether the report has what they want.

Besides being so useful to the reader, abstracts are often lifted out of the report and used elsewhere. For instance, abstracts are used in library files, in abstracting journals and in company files. Many more readers will see the abstract than ever read the report, so they judge the whole report by the abstract. Because an abstract often substitutes for the whole report, it must be complete, accurate and well-written, as well as brief. You can see why it is usually the most carefully written page.

The two kinds of abstracts being used today differ in their purpose and thus in their contents. The shorter version is called a "descriptive abstract," an "indicative abstract" or more often, just "abstract," while the longer version is called an "informative abstract," or an "executive summary," or simply an "abstract."

A DESCRIPTIVE ABSTRACT *describes the contents of a non-decision-making report.*

AN INFORMATIVE ABSTRACT *summarizes the content, conclusions and recommendations of a decision-making report.*

5.6.1 DESCRIPTIVE ABSTRACTS.

Readers look at this kind of abstract to see whether they should read the report, that is, whether it has the information they need. In fact, a descriptive abstract has only one purpose: to describe the contents of a report.

Some writers condense such an abstract into one sentence, but that is the extreme. I recommend that the writer limit the abstract to one paragraph instead. It may help to think of a descriptive abstract as a written form of the table of contents. The captions you have used in the report or the topic sentences you have written for key paragraphs in the report may form the basis for this abstract. A descriptive abstract does not explain ideas or discuss the subject; it simply describes the coverage or scope of the report.

If the report makes no decisions, a descriptive abstract is quite appropriate. Figure 16 is an example of a descriptive abstract written for *Warping,* which, you will remember, is a manual (a non-decision-making report).

5.6.2 INFORMATIVE ABSTRACTS.

An informative abstract (also called an introductory summary or an executive summary) is the longer version of an abstract. It has two purposes:

- to summarize the whole report and
- to reveal its main ideas and conclusions.

Many readers use abstracts as a quick, easy way to stay abreast of their fields. Reading abstracts is an efficient way to absorb a great many research findings in a short time. Some also use informative abstracts just as they do descriptive abstracts – to decide whether to read the report.

ABSTRACT

Warping with a Sectional Beam tells how sectional beaming works, shows

the equipment, explains how to adapt standard warp formulas to sectional

beaming, and gives instructions for warping. The first section describes the

process and the mechanism. Illustrations of the warping in progress, a sec-

tional beam, a horizontal spool rack, and a tensioner are included. Section 2

gives instructions for the process, explaining how to use sectional warp for-

mulas (including formulas for plain weave, twill, twill derivatives and unit

class weaves), how to operate the equipment and how to do the warping. Five

photographs illustrate the process. An appendix lists manufacturers of sec-

tional beams.

v

The first sentence summarizes the entire report.

The title of the report should appear at or near the beginning of your abstract.

Mention special features.

Let the sentence structure reflect the sections of the report.

With a descriptive abstract like this, researchers and cataloguers can see what is covered in the report, and key words can be entered in databases to help future researchers.

FIGURE 16 Descriptive Abstract

An informative abstract presents the conclusions reached in the report but reserves most of the reasoning and details that led to the conclusions for the report itself. Even though the limits are somewhat more flexible, this type of abstract should still fit on one page.

✔ Unless your professor or editor tells you otherwise, decide which type of abstract to use based on whether your report makes decisions.

34 *Formal Reports: Front Matter*

This page may be titled "ABSTRACT" or "EXECUTIVE SUMMARY."

Write the title of the report below the title of the page.

The first sentence encapsulates the entire report. Try to "load" this sentence with the keywords of the report.

The following sentences reveal the report's main ideas and tell briefly what was concluded about each idea.

Just like the descriptive abstract, an informative abstract must describe accurately the scope of the report, which sometimes includes mentioning what the report does not cover.

If you include both a descriptive abstract and an informative abstract, the descriptive abstract comes first. Some professors and editors even want a short descriptive abstract to appear on the title page.

EXECUTIVE SUMMARY
Market Expansion via Internet
A Preliminary Feasibility Study

Six marketing issues need review before adding marketing via the Internet to a standard mix: the Internet's international reach and low airtime costs are advantages while its downward price pressure, poor presentation, lack of distribution channels and lack of market segment control present challenges. For most products, successful marketing on the Internet requires support in more traditional media. Because the Internet mimics television and direct mail, companies considering a rollout via Internet should consult existing marketing studies for specific industries and/or products dealing with those mediums. A bibliography of general studies on these issues is included as an appendix.

v

FIGURE 17 Informative Abstract

From the Editor:
Descriptive vs. Informative Abstracts

Students should consult their instructor about which type of abstract to write. In general, the descriptive abstract is rather limited in its usefulness and is appropriate only for non-decision-making reports. If you have any doubt, write an informative abstract.

Writers on the job would be well advised to write both a descriptive and an informative abstract, since each might be needed when the report is entered into a library's catalogue.

– AMB

5.7 Glossary.

A glossary defines terms as they are used in the report. Definitions can be put in five places, depending on what type of definitions your reader needs and how many definitions there are. In only one case should you use a glossary.

- If you have a few (two or three) very short definitions, put them in the introduction or discussion where the term is first used.
- Once or twice in a report of, say, twenty pages, you can use an informational footnote for a definition of short or moderate length.
- If you have one very long definition, you can put it in an appendix by itself.
- If a long definition is central to the report, you can put it in a paragraph in the introduction.
- When you have three or more definitions of short or moderate length, use a glossary.

A glossary, which may appear in the front matter or as an appendix, may list definitions for words, phrases and abbreviations.

Regardless of where you put definitions, define whenever you know or suspect the word is beyond the technical knowledge of your audience, whenever you are using the word in a specialized way and whenever the reader may be confused because the term is used differently in different technologies. Although a word may have many definitions, give only the meaning used in your report. When you define an abbreviation in the body, spell it out the first time you use it and put its abbreviation in parenthesis immediately after.

The first time you use a defined word in the report, put "see glossary" in parentheses immediately after the word.

5.8 List of Symbols.

If your audience has less technical expertise than you or comes from another discipline, you may need to list symbols. For example, a chemist may list Greek letters and their meanings not because the audience would not recognize them, but because they are used differently in the various sciences and technologies.

Symbols such as Greek letters can be listed alphabetically, but if the symbols have no inherent order, list them in the order they appear in the report. When more than one type of symbol is used, each type is listed separately. Symbols are not listed in a glossary, but if you have just a few, they can be explained parenthetically where first used or in an informational footnote.

Many symbols are now available as computerized characters, but if this is not so, neatly drawn symbols are just as acceptable.

36 *Formal Reports: Front Matter*

Begin definitions of a verb with "to."

A word being used as a word is put in italics or quotation marks. If you have no access to italics, underlining is used in their place.

Don't capitalize just because a word is being defined. Capitalize proper nouns only.

Between the word and its definition, the word "is" or "means" is assumed.

Use any method of definition, such as giving a sample sentence or identifying the part of speech.

GLOSSARY

beam: to turn the warp beam to wind warp on it.

bout: the group of warp ends that fits between two pegs on a sectional beam.

collector: part of a tensioner: a plate with holes drilled in it.

creel: a rack for spools. A synonym for "spool rack" that suggests a horizontal form.

epi: ends per inch; also *sett.*

quill: paper wound in a tight cylinder for use as a bobbin.

sett: ends per inch or epi. Used as a noun, this is a synonym for epi: "It has a sett of twenty." *Sett* can also be used with other units of measure: "The sett is twelve per centimeter." Also used as a verb: "Sett the warp at six."

take up: the warp that goes up over and down under the weft and thereby reduces the length of the cloth; take up forms the depth of the cloth.

tie on: warp tied to the warp beam or cloth beam. This part cannot be woven.

vii

FIGURE 18 Glossary

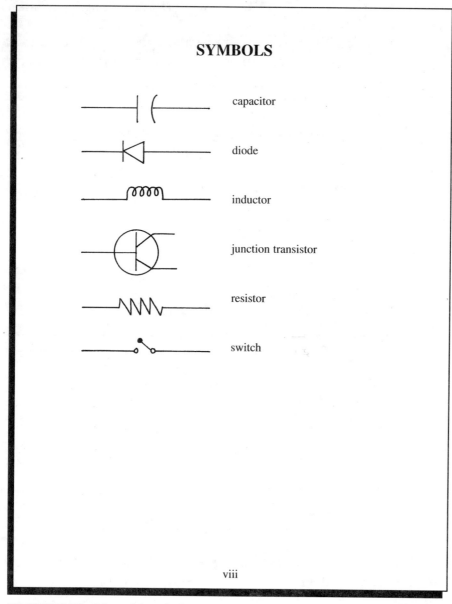

FIGURE 19 List of Symbols

PLAN YOUR REPORT: DRAWING A MAP

It doesn't matter how carefully you follow directions, if you use the wrong map. The check marks in the text should alert you to controversial choices, and they are repeated here to help you create the map your instructor requires for your project.

1. Does your professor require a traditional or decimal outline? _____

2. Does your professor require a different presentation format from that described in section 4.4? _____
_____
_____

3. Which type of transmittal correspondence does your professor require, a memo or a letter? _____ If it is your choice, see section 5.1.

4. Does your professor require a long or a short table of contents? If there is no requirement, see section 5.4 for suggestions about which to use. _____

5. Does your professor require illustrations? _____ If so, how many? How many illustrations may a report have before a list of illustrations is required? _____

6. Does your professor require or prefer a descriptive or an informative abstract? If it is your choice, see sections 5.6.1 and 5.6.2. _____

7. Does your professor have type handling requirements or preferences? _____

_____
If not, see the note below Figure 20.

8. Does your professor require informational or title captions for figures? _____

_____
Is there a requirement regarding numbering and caption placement for tables?

_____
If it is your choice, see section 6.2.1.

9. Is a subhead hierarchy other than the one recommended in section 6.2.2 required or allowed? _____

10. Does your professor require a conclusion in a non-decision-making report? _____ If not, see section 6.3 for advice.

11. Does your professor require a minimum number of quotations?

_____

12. Are both direct quotations and paraphrases required? _____

13. Does your professor require a minimum number or a particular mix of sources?

_____

14. Do you need and have permission to use sources older than ten years?

_____

Formal Reports: Body

CHAPTER 6

The body of a report has two, three or four parts.

A non-decision-making report has two parts:

introduction and

discussion

OR three parts:

introduction,

discussion and

conclusion.*

A decision-making report needs four parts:

introduction,

discussion,

conclusions* and

recommendations.

*The conclusion in a report that does not make decisions and the conclusions section in a decision-making report are quite different in function and content. These differences are explained in section 6.3. For the moment, notice that one has an "s" and the other does not.

This chapter shows an introduction from *Warping,* seven discussion pages – five from *Warping* and two from *Marketing* – and conclusions and recommendations pages from *Marketing.*

6.1 INTRODUCTION.

The introduction prepares the reader to read the discussion. To do this it states the purpose, coverage (or scope) and intended audience of the report. The paragraph telling what the report covers is organized to reflect the organization of the discussion section of the report. Background information, such as a brief history or theory, can also be put in the introduction. This is also the place to make qualifications or to explain the limits of your research. Almost any information the reader needs before reading the discussion may be put in the introduction, but only statements of subject, purpose and scope are required.

39

Although the introduction contains some the same information you have already written in the letter of transmittal or abstract, it differs both in purpose and content. Table I may help you understand how these parts of the report differ and how they work together.

TABLE I
How Transmittal Correspondence, Abstracts and Introductions Differ

	transmittal correspondence	descriptive abstract	informative abstract	introduction
AUDIENCE				
• written to the one person or group to whom it is delivered	✗			
• written to all possible readers, including those who will and those who will not read the report itself		✗	✗	
• written only to readers who intend to read the discussion				✗
WHEN READ				
• read ahead of time*	✗	✗	✗	
• read just before reading the report				✗
CONTENT				
• talks about the report	✗	✗		✗
• talks about the subject and ideas**		✗	✗	
LENGTH				
• restricted to one page	✗	✗	✗	
• no length limit; typically 10%-20% of the body***				✗

*Even those readers of the transmittal correspondence and abstract who intend to continue will probably read the front matter when they receive the report and the rest of the report later.

**The introduction is meant to lead the reader smoothly into the discussion section, while the abstract and letter of transmittal must stand on their own, making complete sense when read separately.

***An introduction for a report of decisions is usually longer than one for a non-decision-making report, because it must introduce ideas as well as the subject.

The introduction in Figure 20 is for the manual, *Warping with a Sectional Beam*. This introduction describes, but in a report of decisions, results would be given. The quotation in it is unusual. An introduction usually has no detail specific enough to require citing, but there is nothing wrong with quoting and citing in an introduction if you have reason to do so.

INTRODUCTION

This report gives instruction for putting a warp on a loom with a sectional beam. The instructions include all the information an experienced weaver needs to adapt from standard beaming. First it describes the beaming process, sectional warp formulas and the equipment. Then it gives instructions for the entire operation. Threading the heddles and sleying are not covered. The equipment shown is typical and the instructions apply to any sectional beam.

Sectional beaming can increase efficiency greatly. One person, working alone, can put on a warp of any length or width. Tension in long and/or wide warps is easier to control. Measuring and beaming are done in one step, not two. If the yarn or thread is bought on spools, beaming can be done in as little as 1/20 the time required when using a warping board or tree and a plain beam. These, plus some minor advantages, make the equipment purchases, the set-up time and the training worthwhile.

Warping is sometimes considered, as Ruth Sisler says, "the least interesting and most difficult step in the textile process," but it is also half the making of cloth.(1:82) Since warping is the first half, the excellence of the whole depends on it.

1

"Gives instructions" states the purpose. The rest of the sentence states the subject.

Sentence 2 identifies the audience as experienced.

Sentences 3 & 4 show the report's coverage and organization.

Sentence 5 qualifies the report's coverage, and 6 states its application.

Paragraph 2 shows the importance of the information.

Paragraph 3 states the importance in another way.

Notice how a partial sentence is quoted and worked into the writer's sentence.

The page number i is always omitted, and the page number 1 is sometimes omitted.

FIGURE 20   Introduction

Notice that the main body of the report is double-spaced with indented paragraphs. Use either Courier, if you are using a typewriter, or a standard serif typeface like Times New Roman for the body. Avoid sans serif typefaces for the body, because they are harder to read.

✔Your professor or editor may have other type handling preferences or requirements. Inquire.

6.2 DISCUSSION.

The discussion section is a report's longest part. Figures 21-25 show sample pages of the discussion section of *Warping with a Sectional Beam*. Four sample pages from *Marketing via Internet: A Preliminary Feasibility Study* are shown in Figures 26-29. Comments beside these figures explain ways to handle the many details in a discussion section.

6.2.1 ILLUSTRATIONS. A problem peculiar to technical reports is the use of illustrations. Technical reports use whatever illustrations can help explain the subject or idea. These include photographs, drawings, tables, charts, diagrams and others. Many fields have developed highly specialized graphics, such as blueprints, circuit diagrams or flow charts, which non-technical readers may need help to understand. Illustrations used to entertain, such as cartoons, are inappropriate in technical reports.

6.2.1.1 Treating Tables and Figures Differently. Tables and figures are both illustrations, as section 5.5 of the Guide explains. However, in several ways, they are treated differently when they appear in a report, as Table II shows. Tables are usually given Roman numerals with a centered caption. Sometimes tables are captioned flush left and/or given an Arabic numeral, though the second change is rarely done when figures also appear in a report. Tables are always given a title caption, which is

TABLE II
How Treatments of Tables and Figures Differ

	tables	figures
Captioned above	✗	
Captioned below		✗
Captioned flush left	✗*	✗
Captioned centered	✗	✗*
Numbered with Arabic numerals	✗**	✗
Numbered with Roman numerals	✗	
Used with a title caption	✗	✗
Used with an informational caption		✗

*Used less often
**Rarely used in a report with figures.

placed above the table. Figures, on the other hand, are numbered with Arabic

> *From the Editor:* How to Identify a Table.
>
> Most figures are easy to identify. Photographs, line drawings, circuit drawings, blueprints and flow charts are obviously figures. However, sometimes it is not easy to distinguish a figure from a table. In that case, use this definition as a guideline: if the information is organized into horizontal rows and vertical columns, it is a table.
>
> – AMB

numerals, captioned flush left below the figure and may be given either a title or an informational caption.

> "Visual information... from a source needs a citation just as much as verbal information."

6.2.1.2 Documenting Illustrations. Illustrations can be drawn or constructed by the report writer, or they can be taken from sources by photocopying, scanning or downloading. Of course, if an illustration is from a source, credit must be given to the source with a citation. Visual information that comes from a source needs a citation just as much as verbal information does.

If the illustration is taken as is, a citation is placed at the end of the caption. If you create your own illustration using information from a source or otherwise adapt an illustration for use in your report, use an informational footnote like the one on the next page of this textbook. Only when an illustration is your own concept and your own creation does it require no citation.

6.2.1.3 Placing Illustrations. Illustrations are not just dropped into a report. Like quotations, they must be blended with your writing. How illustrations are integrated into the report depends on your purpose.

If you think an illustration needs just a glance, a few words to introduce it and a comment after will do. However, most illustrations in technical reports need to be studied, so you must point out what to notice. This is easiest to do if the illustration is placed as soon as possible after it is first mentioned. Once the reader has seen the illustration, continue discussing it until its content is clear. Refer your reader to the illustration by number, and on first mention, give some indication of its content.

> ✓ Does your professor or editor require informational or title captions for figures? Is there a requirement regarding numbering and caption placement for tables? Your professor or editor may leave these decisions up to you.

6.2.1.4 Captioning Illustrations. Tables are given title captions, while figures can be given title captions or informational captions. Informational captions can be as short as a phrase or as long as several sentences. Notice that an informational caption starts with a title, which is often set off by capitals, boldface or other special type use.

This *Guide* shows captions centered above tables and flush left below figures, since that is the usual practice. Allow about a double-space between an illustration and its caption and either three spaces or two double-spaces between it and the text. Capitalize the caption title as you would any title: first, last and all other important words.

6.2.2 SUBHEADS. A second problem technical writers face is the use of subheads. So many type handling solutions are now available that a technical writer today needs to be more careful than ever to choose wisely. A technical report needs a simple, consistent hierarchy of subheads that is appropriate to the seriousness of the subject matter. Subheads must also be distinguished from the captions used for illustrations. Here are three ways to keep your subheads under control:

- Establish a simple, clear hierarchy like the ones shown below and in Figures 21-27. This is one time in your life when you should resist being clever.
- Use your hierarchy consistently. The "cut and paste" or "paint text" functions of your word processor may help you "paint" each level consistently.
- Proof your entire report in several waves, reading through only for sentence structure, only for spelling, only for capitalization, etc., and be sure you use one of those waves to look only at subhead type handling.[a]

✓ Before you use any other subhead hierarchy, check with your professor or editor.

Here are instructions for establishing a clear hierarchy among your report's subheads: Begin the discussion section of your report by centering and boldfacing the word **DISCUSSION** $1\frac{1}{2}$" down from the top of the page in all caps. Triple space, indent and begin your first paragraph OR triple space and type the first subheading. Here, one heading may follow another with no intervening text. Elsewhere, avoid stacked headings. Because you should start a new page whenever you use a first level heading, these headings always appear at the top of the page.

Write second level headings in all caps and boldface. Place them on the left margin (flush left). A second level heading looks like this:

1 A SECOND LEVEL HEADING.

Double space down, indent and begin a new paragraph. Third level headings are also flush left and boldface, but they are in upper and lower (U&L) case, so they look like this:

1.1 A Third Level Heading.

After a third level heading, double-space down, indent and begin a new paragraph. Third or fourth level headings are called paragraph headings when they head a single paragraph. Fourth level headings are indented, written in upper and lower case like a title, boldfaced and ended with a period. A fourth level heading looks like this:

1.1.1 A Fourth Level Heading. Begin the paragraph that follows on the same line.

---

[a]Notice that each item on this list is one or more complete sentences and that it is introduced by a complete sentence. This is one of three ways lists can be treated. The alternatives are to treat the entire list as one sentence or to treat each item on the list as a separate ending for the sentence above (see Figures 21 and 22).

Formal Reports: Body **45**

If your report needs more than four levels, consult your instructor or editor for a more complex hierarchy of subheads. This system of subheads assumes you have boldface and italics available, but it also works with neither.

6.2.3 SAMPLE DISCUSSION PAGES. Figures 21-24, from *Warping,* show a fairly complex set of headings, two illustrations, three citations and an informational footnote. They are followed by two sample discussion pages from *Marketing,* where you will notice a change from description and instructions to analysis of ideas.

DISCUSSION

1 THEORY OF OPERATION.

The basic process of sectional warping takes place in three steps:

- The warp amount is calculated, and the warp is wound on spools on a rack.

- The ends are threaded through a tensioner and tied to the warp beam.

- The beam is turned, and the yarn rolls off the spools and onto the beam.

The whole width of the warp cannot be wound on at once without a spool for each end and a huge spool rack and tensioner to accommodate them. (Mechanized looms do use such equipment.) Instead, the number of spools corresponds to the number of warp ends in two inches, and each two-inch space between pegs is wound on one at a time. When one 2" bout (See Glossary.) has been wound on, the rack and tensioner are moved sideways and the next bout is put on. This process is repeated for the whole width of the warp.

1.1 Adapting Standard Warp Formulas.

To buy and prepare warp for sectional warping, you must know this information:

- the number of warp ends per inch,

- the width of the warp in the reed,

2

You may number the items on a list or use bullets.

Notice the whole sentence in parentheses.

Write out short numbers except where it might be confusing.

Parts of the report are capitalized. Allow 3 spaces above a heading.

This is a third level heading: flush left, U&L case and boldfaced. Start the next sentence on a new line.

This whole list is one sentence. Compare to the first list on this page, which uses one sentence per bullet.

FIGURE 21 *Warping,* Discussion, Page 2

46 *Formal Reports: Body*

The paragraph continues after the list, so it is not indented.

This is a fourth level heading: indented, U&L case and bold-faced. Start the next sentence on the same line.

Notice the words are spelled out *before* the first use of an abbreviation.

Either write out numbers and symbols or use the short form of both: two inches or 2".

Formulas are centered on their own line unless they are very short.

• the length of the warp,

• the total yardage,

• the number of spools, and

• the length of warp to be put on each spool.

The first three are found by deciding the density, length and width of the warp. The number of spools equals the number of ends in the space between each pair of pegs. Typically, pegs are set at two inches, so the number of ends in two inches determines the number of spools.

1.1.1 Calculating Total Yardage. Yardage must be calculated before a weaver can buy materials. First, multiply the ends per inch (epi) times the warp width to find the total number of ends in the warp. Multiply this product by the length of the warp. For example, to weave 24 yards of 40" wide fabric at an epi of 16,

$$(16 \text{ epi x } 40") \text{ x } 24 \text{ yds} = 15,360 \text{ yds}.$$

This formula will give you the minimum amount of warp yardage needed.

1.1.2 Calculating Yardage per Spool. The total yardage must be split equally among the number of spools used per bout. To find the yardage per spool, first determine the number of spools by multiplying the epi times the number of inches between pegs. With an epi of 16 and two inches between

3

FIGURE 22 *Warping*, Discussion, Page 3

Third and fourth level headings are capped as titles are: first, last and all other important words. The alternative is initial caps, which are easier (no worry about whether a word is "important") but are considered sloppy by some readers. Ask your instructor or editor whether he or she allows initial capping. If you have any doubt, use title capping.

pegs, the weaver needs 32 spools. Since each spool must carry 1/32 of the warp or 15,360 yards, each will need 480 yards. The whole formula is

(epi x width)length = yardage ÷ number of spools = yards per spool

or in this example,

(16 epi x 40")24 yds = 15,360 yds ÷ 32 spools = 480 yds/spool.

The weaver buys 15,360 yards, optimally on 32 spools of 480 yards each or the closest higher increment, probably 500 yards.

You may occasionally find yarn in skeins with no yardage marked, since some manufacturers specify weight only. If you know a yarn or thread's type, weight and ply, you can calculate its length. Ruth Sisler's excellent instructions in *The Weaving Handbook* have been reproduced with permission as Appendix A of this report.

If a warp is to be in stripes or in uneven weaves, proportions of each must be figured separately. Tables showing warp length and yardage for typical products are shown in section 2.1, "Using Sectional Warp Formulas."

1.2 Using the Mechanism.

Warping with a sectional beam requires a spool rack, a tensioner and a bobbin or spool winder. For good spool winding instructions, refer to Amy

4

> **Normal punctuation is used before and after a formula.**

> **Be sure to mention an appendix at the appropriate place in the report.**

> **The name of a part of a report is written in caps with no quotation marks, but the headings are titles, so they are written in U&L case with quotation marks. Compare Appendix A to "Using Sectional Warp Formulas."**

FIGURE 23 *Warping,* Discussion, Page 4

The last paragraph on this sample page introduces section 1.2, explains why spool winders are not discussed here and tells the reader where to find more information. It also introduces Figure 1.

Always introduce an illustration before it appears.

This is an informational caption. Boldface the number and title and single-space the sentence(s) that follows.

The caption contains a paraphrase.

Allow 2 spaces after an illustration.

This is an informational footnote. Notice the superscript letter that connects it to the text.

Hoover's directions published in the winter 1997 issue of *Shuttle & Spindle*.[a] Because most weavers are already familiar with bobbin/spool winders, they are not discussed. Figure 1 is a line drawing from Sinclair Dunlap's Advanced Weaving *Techniques* that shows the essential equipment for sectional warping.

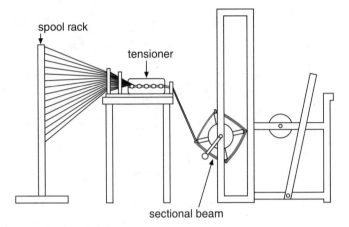

FIGURE 1 Sectional Beaming with Warping in Progress.(2:26) The tensioner can be attached to the loom frame, as shown here, or to a table behind the loom. Svenson recommends 5' between the spool rack and the tensioner and 3' between the tensioner and the beam.(3:112)

In addition to the equipment shown above, a counter is often attached between the tensioner and the beam.

---
[a] Reprints of Hoover's article are available from The Weaver's Source, P.O. Box 1506, Yuma, AZ 54321. It appeared originally as "BobbinWinding" in *Shuttle & Spindle,* vol. 156, winter 1997, pp. 39-42.

5

FIGURE 24 *Warping,* Discussion, Page 5

In this example, the illustration came from a source, so it is cited. Since it is the first citation in the report, and it appeared on page 44 in the source, the citation is written as (1:44). The information in the caption came from a different source and requires the second citation. Notice that the authors who gave this advice are mentioned by name. This is called an attribution, and it should be used whenever possible, so the reader understands that you are moving from what you know from experience to what you read or heard from a source.

1.2.1 The Sectional Beam. The beam itself should be very strong with wood dowels or metal pins (about 1/4" diameter) every 2" to separate the sections. Figure 2 from Svenson's *Handweaving for Professionals* shows a

FIGURE 2 A Sectional Beam.(3:116) Notice the tape securing the completed bouts and a temporary weft maintaining the cross established in the tensioner.

sectional beam with the fourth bout being wound on. This illustration shows warp starting at one end of the warp beam, but if a warp is narrow, it is centered.

1.2.2 The Tensioner. All tensioners establish an even tension among warp ends, but, as Dunlap points out, "more highly developed tensioners have advantages."(2:28) The tensioner he recommends, shown in Figure 3, has a friction drum and a variably weighted brake. As each thread comes from the

6

> The writer knew this from experience, so it is not cited.
>
> If you write informational captions, write one for every figure. Tables are not given informational captions.
>
> When the same book by Svenson is cited again, it retains the number 3.
>
> Compare this direct quotation to the paraphrase in the caption for Figure 1.

FIGURE 25 *Warping,* Discussion, Page 6

The next two sample discussion pages are taken from the report on Internet marketing. They show a typical mix of quotations, paraphrases and original writing, as well as several types of quotations. Notice how the quoted material is blended into the writer's own words. The point of the report is to identify areas needing more study by manufacturing or distribution companies considering marketing their products via Internet.

Because *Marketing* is a decision-making report, it analyzes, using source material to explain and support its conclusions.

50 *Formal Reports: Body*

This direct quotation picks up in the middle of a sentence and then includes another whole sentence. The next sentence quotes a key term and a phrase. These are ways to incorporate words from your source into your own sentences smoothly.

Paraphrase statistics and facts. Quote opinions directly. Compare with the first paragraph.

Mixing more than one source per paragraph demonstrates that you are not letting your sources do your thinking for you.

Use a title (like *American Demographics*) only when no author's name is given.

Underlining and italics are interchangeable. Use italics if you can.

of any entrepreneur, but as Lester Wunderman reminds us, reaching the consumer directly is not about "an ad with a coupon, a catalog, a phone call, a database or a web site. It's a commitment to getting and keeping valuable customers."(3:52) Martalo calls the Internet a "tactic" and urges us to "think first of selling strategies."(4:133) A company can prepare for digital marketing by cross-referencing existing marketing statistics with Internet usage figures.

The percentage of homes with PCs creeps up, but resistance to using computers at home is strong. An American Society for Quality/Gallup poll cited difficulty of use, high cost and low interest as barriers to computer use.(5:35) The number of homes with PCs grew by only 2% per 6 months in 1997: 37% in January 1997, 39% in July 1997, 41% in December 1997, according to *American Demographics*. (2:13) At this rate, the 100% mark will not be reached until 2007. It will be another decade before "everyone" uses a computer at home.

A. Cross-Referencing to Internet User Data.

Key user data include gender, income/education and age. As of December, 1997, Conger and Jones report that 64% of Internet users were male and the median age (though rising rapidly) was 32.(6:296)

3

FIGURE 26 *Marketing*, Discussion, Page 3

The middle paragraph on this page is a great example of how to mix sources and of how to put enough of your own words into a documented report. It begins with a topic sentence written by the student. Next comes a paraphrase – information from a source, but in the student's words. Another paraphrase from a different source follows. The next sentence is not cited because the student writer created this statistic. The student's own conclusion completes the paragraph.

Matthew Grimes lists median income as $57,000 and the typical education level as "college degree plus some graduate work." (7:115) In certain product categories, suppliers may need to consider more specific user profile characteristics.

B. Geographical Definition on the Internet.

One of the undeniable attractions of the Internet is its immensity. The Internet is, as *Fast Company* says, "an electronic global village...the world's biggest stage" with an "audience of millions." (8:269) But that very size causes several serious problems.

Many companies are not able to take advantage of the international nature of the Internet. Shipments of food, for example, are heavily regulated. Most furniture is too heavy for cost-effective long-distance shipping. Many products can be efficiently sent only via cargo ship container, and that requires huge quantities, not individual sales. To the extent that worldwide marketing is impractical for these companies, the dollars they allocate to Internet marketing are wasted.

The Internet also weakens traditional distribution channels. Consumers may, in fact, need middlemen in some cases. Retailers provide a setting

4

> This paragraph is another good mix of sources controlled by a topic sentence and conclusion in the student's own words. Notice the direct quotation to let us know exactly what language was used on the survey quoted.

> This paragraph introduces an analysis of a problem. Citations often decrease in areas of analysis, which depend on the writer's own thinking. Citations may well be spread unevenly though the report.

FIGURE 27 *Marketing,* Discussion, Page 4

In decision-making reports, ideas are discussed and conclusions are reached right in the body of the report. These conclusions are then gathered and repeated in one section of the report, the Conclusions page. This means you should not "save" your conclusions for the Conclusions page. Instead, let them first appear as they naturally occur in the discussion section.

6.3 CONCLUSIONS.

Conclusions and recommendations sections are included in decision-making reports only. Such reports have actually reached conclusions on which to base recommendations. Some instructors or editors require a conclusion (note that in this case the word is not plural) in non-decision-making reports, but since it serves little purpose except for waving a polite good-bye, I cannot endorse it.

Rather, I suggest the conclusions section serve the important purpose of collecting on one page all the conclusions reached in the report. A conclusions page should be complete, listing all conclusions, but this is not the place to explain or defend. The evidence that led to a conclusion should be given in the discussion section where the conclusion was originally stated. Instead, the conclusions section says nothing new; it simply isolates the conclusions so they can be seen at one glance.

A conclusions page may be written in paragraph form or as a list. Sometimes more than one list is used, grouping, for example, experimental results and theoretical conclusions. Whatever the form, the conclusions and recommendations pages should work in tandem. By showing all the conclusions right before the recommendations, you make the connections more obvious.

Matching your conclusions page to other parts of the report may make it easier to write. Present conclusions in the same order in which they appeared in the discussion. The content of the conclusions and recommendations pages, taken together, should match the informative abstract closely. A sample conclusions page from *Marketing* appears on page 53 as Figure 28.

✔ **Ask whether your professor requires a conclusion on a non-decision-making report.**

6.4 RECOMMENDATIONS.

Some writers feel they do not know enough to make recommendations, but it must be done. A decision-making report's whole purpose is to present and analyze information so a decision can be made and action taken. If you have studied the issues thoroughly enough to write about them, you must have formed opinions about a sensible course of action. If you state recommendations carefully, with whatever qualifications are needed, you may not feel so much as if you are putting yourself out on a limb. If you double-check to be sure the recommendations are backed by the conclusions and the conclusions are backed by the evidence and reasoning in the discussion, you will not be limb-walking at all.

Remember, too, that the executive deciding the issue will make up his or her own mind. Your recommendations are only as valuable as your research and reasoning, and even if that is faultless, an executive may decide differently based on information you could not know. Realize how important recommendations are and write them carefully, but do not be personally hurt if they are not followed.

Sometimes one recommendation follows from many conclusions, but just as often, each recommendation corresponds to a conclusion. For this reason and others, a recommendations section can be written as a paragraph, a list or a combination. You will find a sample recommendations page from *Marketing* as Figure 29 on page 54.

```
                    CONCLUSIONS


     Marketing through the Internet is clearly in its
infancy, and a clear picture of opportunities and
pitfalls is still emerging.  Remarkably low airtime
costs and a truly worldwide reach have led to a
number of spectacular successes.  Companies with
established brands, however, must exercise care to
avoid damaging product status, thereby threatening
profits.  The Internet exerts downward price pres-
sure, presents products poorly, and weakens distrib-
ution channels and market segment control signifi-
cantly.  In most product categories, support from
traditional media alerts consumers and increases
desire as well as helping consumers access web
sites and complete a purchase.  The similarity
between television, direct mail and the Internet
makes existing marketing studies of the first two
an appropriate resource for businesses, especially
those with access to reports specific to their
product or industry.




                          16
```

The writer begins with a qualification, and follows it immediately with the two advantages and four disadvantages mentioned in the Executive Summary of this report. (Compare Figure 16, page 33.)

FIGURE 28 *Marketing,* Conclusions

54 *Formal Reports: Body*

This report is a preliminary feasibility study for any company considering using the Internet. Most such reports are written for a specific company and would refer to the company by name.

Match the order of your list of recommendations to the conclusions and to the order in which the issues related to them were covered in the report.

Begin each recommendation with a verb because you are recommending action.

RECOMMENDATIONS

I recommend a company considering marketing its products or services via the Internet undertake these studies:

1. Gather actual, current costs (domain name registration, content development, programming and server fees).

2. Compare competitors' products on the Internet with special attention to price movement for these products since they were brought to the Internet. Use the results of this study to adjust the assumed price reduction percentage needed to implement recommendation number 3.

3. Re-calculate return on investment, assuming downward price pressure of 15% ± and sales growth between 30-100%.

4. Evaluate effects of weaker distribution channels.

5. Evaluate effects of loss of market segmentation.

6. Review the most specific studies available regarding television and direct mail marketing for the company's industry or product category. These studies may be helpful in implementing recommendations 2, 4 and 5.

17

FIGURE 29 *Marketing,* Recommendations

Formal Reports: Back Matter

"All's well that ends well."
– Shakespeare

CHAPTER 7

The three parts of a report behind the body are the references page, appendixes (optional) and a flyleaf. The references page and appendixes are discussed below. The back flyleaf is a blank, unnumbered sheet placed just inside the back cover.

7.1 LIST OF REFERENCES.

A references page lists the sources you have used in your report. Sources are listed in the order in which they were cited in the report. Sources you read but did not cite can be listed without numbers below a numbered list of those you did cite. The example shown in Figure 30 is from *Marketing via the Internet*.

7.2 APPENDIXES.

A report's appendix, like a person's, is unnecessary. It may be helpful, but it is not essential. These types of information may be appended:
- balance sheets
- calculations, especially in long form
- case studies
- correspondence
- leaflets or brochures
- a list of suggested readings

In fact, any information that is nonessential to the report can be appended. Appending material frees the report from distractions. However, do not move anything to an appendix that most readers will need to make sense of the discussion.

Appendixes can consist of so many things that it is hard to generalize about an appropriate format. However, two useful formats are described here. Most short appendixes can be typed on a page with a heading placed $1\frac{1}{2}''$ from the top of the page. Such a heading would look like this:

APPENDIX
ANNOTATED LIST OF SUPPLIERS OF SECTIONAL BEAM LOOMS

56 *Formal Reports: Back Matter*

An appendix that cannot be handled this way, such as a plastic pocket for slides, can be inserted with a title page before it. Put the same kind of heading about 3½″ down on such a title page. If your report has more than one appendix, label them A, B, C, etc. Continue the same page numbering you have been using for the discussion and references for the appendix pages.

REFERENCES

This book with 3 authors was the first source cited in the report.

1. Stull, Milton, and others, *Handbook for Advertising Managers,* 3rd ed., Englewood Cliffs, NJ, Prentice Hall, 1997, pp.62-85.

A journal article, no author

2. "PC Market Sluggish," *American Demographics,* vol. 3, no. 6, June, 1999, p. 13.

A book, 1 author. Notice the subtitle.

3. Wunderman, Lester, *Being Direct: Making Advertising Pay,* New York, Random House, 1996.

A journal article, 1 author

4. Martalo, Josephine, "Are You Ready for the Internet?" *American Demographics,* vol. 3, no. 4, April, 1999, pp. 131-147.

A book, no author listed

5. *Computer Industry Trends,* New York, Dreser-Norstrand Institute of Research, 1999, pp. 28-37.

A book, 2 authors

6. Conger, John A. and Linda M. Jones, *Influential Americans: Marketing to the Trendsetters,* New York, Information Data Publications, Inc., 1998.

A book, 1 author

7. Grimes, Michelle, *The Right Site,* Naperville, IL, Thinkbank Publishers, 1997.

A magazine article, no author

8. "The Web Can Make You a Star!" *Fast Company,* January, 1998, pp. 268-270.

A book, 1 author

9. Lewis, H.G., interviewed by the author, The University of Akron, Akron, OH, 2 PM, Feb. 5, 1999.

An interview by the report's student writer

Garfinkel, Simson, "Web Brownout," *Wired,* vol. 6, no. 9, September, 1998, pp. 94-99.

These three sources were consulted, but not cited. (No quotation, paraphrase or illustration was used from them.) Such sources are unnumbered and alphabetized.

Grimes, Michelle, *The Virtual Company,* Naperville, IL, Thinkbank Publishers, 1999.

Walker, Rebecca and Angel Williams, *Old Ways, New Ways,* San Diego, Humanitarian Press, 1999, p. 6.

18

FIGURE 30 *Marketing,* References

✓ Does your professor or editor require a
minimum number of sources?
Is a certain mix of sources required?

Documentation

document: *tr.v.* 1. to support with written references or citations. [from Latin *documentum*, lesson, from *docere*, to teach.]

CHAPTER 8

Welcome to the numbered reference style of documentation. Numbered reference documentation uses simple,[a] consistent forms with a close resemblance to those recommended by the Council of Biology Editors. Not only are you likely to appreciate the simplicity of this style, but it will also prepare you to use more complex styles. Regardless of your personal preferences, you must always use whatever documentation style the editor or instructor requires when you are preparing material for class or for publication.

More important than format are the purposes and principles of documentation, which remain the same in all styles. Documentation always functions in basically the same way: it allows your reader to trace your use of sources. While you should pay strict attention to the details of the numbered reference style this semester, be prepared to change to whatever form of documentation your next professor or editor requires.

Documenting thoroughly and accurately is the way a writer of a research paper makes his/her work legitimate. By quoting and paraphrasing, a writer

- gives credit to those who deserve it.
- verifies information not known from firsthand experience.
- shows that information was checked. Even a writer with firsthand knowledge needs to show that others agree.
- demonstrates that a variety of sources were consulted. This establishes that the research was broad enough and that writer did not become over-dependent on a single source.

Attributions and reference page entries provide information about the sources to assure the reader of the sources' credibility.

> **"Documentation ... allows your reader to trace your use of sources."**

---

[a]This is a footnote, or note at the foot of the page. You will be happy to learn that the numbered reference method does not ask you to use footnotes for identifying your sources. Use footnotes only if you need them, as I have here, for extra information.

Attributions and reference page entries
- identify sources by name.
- show that the sources are reasonably recent. Except for historical sources, ten years is generally considered the "age limit" for a source. In some disciplines and industries, ten years is too long, and in a few, it is too short.
- invite the reader to double-check the sources.

✓ **Use no sources older than ten years unless you have other directions or permission from your professor or editor.**

There are four steps to documenting well:
- quoting,
- attributing,
- citing, and
- writing the references page.

Each should be done with absolute accuracy.

8.1 QUOTATIONS.

Quoting can be *direct,* which means word for word, or *indirect,* which is called paraphrasing. Paraphrasing means using your own words to relate information from a source. Both kinds of quoting require attributions, citations and entries on the list of references.

8.1.1 DIRECT QUOTATIONS. A direct quotation must be written exactly as it appears in the original. Only three changes are allowed:

1. You may change a capital letter to a lower case letter when the word no longer begins a sentence.
2. You may change a lower case letter to a capital letter when a word now begins a sentence.
3. You may change a period, colon or semicolon to whichever of those marks is required by your new sentence.

Not even these changes are allowed where they would change the sense of the original.

The examples below are written as if a writer were quoting and paraphrasing from a technical bulletin published by Linden Industries. A reprint of the bulletin appears as Figure 31. Read it first, and then compare the quotations with the original.

To quote a whole sentence:

```
Injection molding requires careful control of temperatures throughout
the process. Linden Industries says "Tank jackets or in-line tube and
shell heat exchangers maintain temperatures, typically at between 90°
and 130° F."(9)
```

REINFORCED REACTION INJECTION MOLDING

A Linden Industries Technical Bulletin

Reinforced Reaction Injection Molding (RRIM) is a method of blending and molding polyol and isocyanate, components of polymers. The polyol is usually loaded with a filler, and a third component is sometimes added. The two or three components are impinged on each other at high pressure inside a mixing head. The pressure is dramatically reduced, and the mixture is injected at low pressure into a closed mold. There, a reaction takes place, producing a finished polymer in the shape of the mold cavity.

Polyol, isocyanate and the third component, if used, are supplied from feed tanks. Tank jackets or in-line tube and shell heat exchangers maintain temperatures, typically at between 90° and 130°F. Polyol tanks usually have an agitator and a nucleation system to lower the density of the liquid by introducing nitrogen.

From the tanks, feed pumps transfer components through supply lines to metering equipment. Metering units measure and dispense components precisely as they pass to the mixing head. To maintain controlled temperatures, viscosity and density, components are recirculated at low pressures between shots.

As injection begins, valves inside the mixing head re-direct the liquid from recirculation to injection and force it, under pressures between 2,000 and 3,000 PSI, into the mixing head. The components are then mixed by high velocity impingement. After mixing, the reacting material flows into the mold cavity.

When an injection cycle is finished, a clean-out piston pushes all reaction material from the mixing chamber into the mold. Because the mixing head is attached to the mold, all reactant material is removed when the molded part is. With total self-cleaning, flushing with solvents is unnecessary, and the problems associated with this have been eliminated.

RRIM requires a precise control of the flow of liquids. Our expertise in this specialty has allowed us to develop highly innovative, exceptionally productive, low maintenance RRIM machinery.

LINDEN

Linden Industries, Inc.
137 Ascot Parkway
Cuyahoga Falls, Ohio 44223
330-928-4064 fax: 330-928-1854

FIGURE 31   Excerpt from *Reinforced Reaction Injection Molding: A Linden Industries Technical Bulletin*[a]

---

[a]Reprinted with permission from *Reinforced Reaction Injection Molding: A Linden Industries Technical Bulletin*, Linden Industries, Inc., Peninsula, Ohio, 1998.

To quote part of a sentence:

```
Injection molding equipment uses a variety of means "to maintain con-
trolled temperature, viscosity and density" at every step of the
process, according to a Linden Industries' technical bulletin on
reinforced reaction injection molding.(9)
```

Notice the word "to" at the beginning of the quoted material has been changed to lower case.

To omit something within a quotation:

An ellipsis is placed where anything (a word, phrase, sentence or punctuation mark) is omitted within a quotation. Do not use an ellipsis at the beginning or end of a quotation, unless you have an unusual need to alert the reader that something is missing. An ellipsis is a series of three dots. When an ellipsis appears at the end a sentence, a fourth dot is used for a period.

```
Linden describes the injection molding process: "valves inside the
mixing head re-direct the liquid from recirculation to injection and
force it ... into the mixing head."(9) This process is used to
```

When an ellipsis runs over the end and/or beginning of a sentence, it includes the period, so it looks like this:

```
"The components are then mixed .... [and] the reacting material flows
into the mold," Linden reports.(9)  Molds, which range from minute to
```

Like the period in this example, other punctuation marks can be retained if needed.

To add something inside a quotation:

Anything added within a quotation is set inside brackets, not parentheses. Brackets and parentheses are two distinct marks of punctuation and are not inter-changeable. If you are not working with a computer or typewriter that gives you brackets, pencil them in. In the example above, the word "and" is added to make grammatical sense. In the sentence below, the word "polymer" is added to explain what "polyol" means.

```
As Linden says, "Reinforced Reaction Injection Molding (RRIM) is a
method of blending and molding...polymers. The polyol [polymer] is
usually loaded with a filler."(9) Other manufacturing methods rely
```

OR

As Linden says, "Reinforced Reaction Injection Molding (RRIM) is a method of blending and molding...polymers. The ... [polymer] is usually loaded with a filler."(9) Other manufacturing methods rely on

These examples of quotations with ellipses and brackets hint at the reasons writers use them. Your sources may go into greater depth than your report does; they may be written for an audience with more technical knowledge. Adapting quoted material to your purpose and audience is an important part of your job as a writer. However, you will do your audience no favor to complicate quotations unnecessarily. For this reason, you should avoid ellipses and brackets; when you must use them, use them with care.

To use a long quotation:

Any direct quotation longer than three typed lines is considered a long quotation. It is single- spaced and double-indented (indented from both margins), so it looks like this:

of the many processes used. Linden Industries explains how pressure is used to mix and mold polymers in the reinforced reaction injection molding process:

> Components are impinged on each other at high pressure inside a mixing head. The pressure is dramatically reduced and the mixture is injected at low pressure into a closed mold. There a reaction takes place, producing a finished polymer in the shape of the mold cavity.(9:6)

Besides the precise control of shifting pressures, McCann points out that temperature must be controlled through every step, starting with

Notice an unusual, but important, convention of long quotations: the quotation marks are omitted, even though it is a direct quotation.

✔ **Check to see whether your professor requires you to use both direct quotations and paraphrases. Ask also about the minimum number of each required.**

8.1.2 PARAPHRASES. Paraphrasing means putting someone else's information into your own words. Since you have an obligation to retain the precise meaning of the original, you should use paraphrasing when factual content makes the phrasing of the original unimportant. The paraphrase below begins with the writer's own topic sentence and supports it with information the writer knew from working at McCann Plastics; a direct quotation from an interview with Michael McCann, president of McCann Plastics; and a paraphrase of information that appeared in Linden's technical bulletin. The conclusion (in the writer's own words) begins a transition to the content of the paragraph to follow.

> Creating and molding polymers for an immense range of products requires precise control of the entire process. At McCann Plastics, the lab constantly runs tests for tensile strength, color fastness, opacity and many other attributes. According to Michael McCann, President of McCann Plastics, "batch testing in a lab is the only way to be sure that the chemicals with these characteristics are evenly distributed" throughout each batch.(8) And, as a Linden Industries technical bulletin points out, sophisticated controls of pressure and temperature are needed to keep liquid polymer at the appropriate density and viscosity before and during molding.(9) Clearly, precise control mechanisms are essential if polymers are to mimic the qualities of other materials.

When you paraphrase, you must use your own words in your own way. Do not borrow the original author's sentence structure or his words. Some people rephrase the original twice – once when doing the research and once when drafting the essay – to make it easier to "escape" from the original phrasing. Even though you are putting this information in your own words, you are borrowing the information itself, so you should give the original author credit with an attribution and citation just as you do with a direct quotation.

When you paraphrase inadequately, you have overused words or phrases from your source. It is a dangerous form of plagiarizing. To your reader it looks lazy – as if you are cutting and pasting, letting the original writer do all the work of finding the right words and phrases and then taking credit for it. Observe the rule carefully: when you omit quotation marks, the language must by yours, even though the information came from your source.

8.1.3 PLAGIARISM. Rearranging the author's words or lifting passages directly without complete, accurate quotation marks and citations is plagiarism. Here is the paraphrase you just read, but plagiarized:

> these characteristics are evenly distributed" throughout each batch.(8) During blending and molding, metering units measure and dispense components precisely, maintaining pressure, temperature, viscosity and density. Clearly, precise control mechanisms are essential

This passage is plagiarized in several ways. First, it gives no credit to Linden as the source of the information. That alone would constitute plagiarism. It also relies too heavily on Linden's phrasing to use them without quotation marks.

Plagiarism is cheating, of course. As such, it undermines the value of the degree received by every honest student. You should never approximate what someone else said. Instead, either quote directly, with the appropriate marks; or paraphrase, changing everything you can. Whichever you do, cite each use of someone else's material and write a bibliography.

Documentation   63

8.1.4 WRENCHING FROM CONTEXT. The plagiarized passage in 8.1.3 is also is an example of another important problem that can occur with quoting: wrenching from context. Wrenching from context damages the quoted material by ripping it carelessly from the original. In this case the careless paraphrase makes it sound as if metering units control pressure, temperature, etc., while the original clearly says they do not. Whenever you quote or paraphrase, you have a responsibility to preserve the meaning of the original.

8.2 ATTRIBUTIONS.

An attribution is a phrase appearing immediately before or after a quotation or paraphrase that identifies the source. Attributions are used to keep readers informed about where information is coming from as they read. Readers should not be required to check the references page for such information, but instead, should have the list of references available as a resource.

By the way, it is no accident that the word "tribute" is buried in "attribution," since that is where "attribution" came from. Giving tribute to the experts just what an attribution does.

You should write an attribution every time you move from your own information to that supplied by a source, unless the context makes clear to the reader that you are quoting or paraphrasing. Besides their central uses of alerting the reader and identifying the source, attributions also can give the reader important information about your sources. Use them to reassure readers about the date of a source, to tell them which university a researcher is associated with, to identify the type of source, etc. You can see many examples of attributions being used this way in this chapter and in the sample discussion pages in Chapter 6.

> "Whenever you quote or paraphrase, you have a responsibility to preserve the meaning of the original."

8.3 PARENTHETIC CITATIONS.

Each time you use a quotation or paraphrase, use an attribution and a parenthetic citation. The citation is placed after the period (and quotation marks, if any) at the end of the sentence. It includes the number of the reference page entry for this source and the page number from which you took the information in the source:

```
Cohen and Stewart define natural laws as "context-dependent
regularities."(3:285)
```

Notice that the two numbers in parenthesis are separated by a colon.
- If your source has no page numbers (for example, an interview, personal letter or lecture) use the number of the reference page entry by itself: (3)
- If you use information from pages 128 through 129, as when the sentence or passage begins on one page and ends on the next, use this form: (6:128-129)

- The form (2:14,27) is used in the rare case when you use information from pages 14 and 27 in the same passage.
- If your source uses a compound number or unusual number, use the same form in your citation: (5:D1) or (7:ix)

Citations refer to the preceding passage, but sometimes it's hard to tell what a "passage" is. Here's a rule of thumb: the preceding sentence is the preceding passage unless there is an obvious connection in grammar or meaning between the preceding two sentences. A passage is only rarely three or more sentences. A long quotation is an exception to the one-or-two-sentences rule. Anything set off as a long quotation is considered a passage, regardless of the number of sentences and requires just one citation.

BIBLIOGRAPHY:

n. 1. a list of works by a specific author or publisher, or of the sources of information on a specific subject. [French *bibliographie*, from New Latin, *bibliographia*, from Greek *bibli*, book (see *Bible*) + Greek *graphein*, to write]

8.4 REFERENCE PAGE ENTRIES.

Information about your sources is collected in a list at the end of your report. The generic term for a such list of sources is "bibliography," but in the sciences and technologies, the bibliography is usually titled REFERENCES or WORKS CITED.

Choose one of those titles (You may want to ask your professor or editor which he or she prefers.), and center it in all caps at the top of the page. Place your sources on the list in the order in which they are first cited in the report. Number the list. Any sources which you read but did not cite should appear at the end of the list, unnumbered and alphabetized.

This is the data you must record for each source:
- author,
- title information and
- publication information.

You will usually find this information on the title page and the back of the title page. Remember that the point of a bibliography is to allow your readers to find the sources for themselves, so you must hunt for the information your readers need to trace your thinking and evidence to its sources.

The examples below show how to write various reference page entries. The entries differ because the information a reader needs to find a source changes with the type of source.

8.4.1 TYPICAL SOURCES.
For a book by one author

```
1. Truman, Kathryn A., Acoustical Engineering: An Introduction to
    Sound and Physics, Boston, Allyn and Bacon, 1999, pp. 115-299.
```

Notice:
- Titles and subtitles are separated by a colon (even if a colon does not appear on the title page).
- The first word of a subtitle is capitalized.

- Titles of books can be underlined or italicized. Italics and underlining are interchangeable in meaning, but italics are preferred if your writing equipment allows them. Be consistent about this choice.
- The place of publication is given as a city unless the state is also needed to identify the city.
- If you read part of a book, list the pages you read.

For a book by two or three authors

2. Cohen, Jack and Ian Stewart, *The Collapse of Chaos: Discovering Simplicity in a Complex World,* New York, Penguin Books, 1994.

Notice:
- Only the first author's name is inverted.
- Listing no page numbers means you read the whole book.

For a book by more than three authors

3. Nakajima, Y., and others, *Handbook of Sensation and Perception in Experimental Psychology,* 7th ed., New York, W. W. Norton & Company, 1998, pp. 103-107.

Notice:
- Compare the way the authors' name are listed in the three examples above and note that you should not shorten an author's name. The choice of whether to use a full name or initials belongs to the author.
- Some writers use the Latin phrase "et al." in place of "and others."

For an article from an edited book

4. Tobolsky, June, "The Relationship of Ecosystems to Sulphur and Selenium," *Inorganic Compounds,* ed. by C.S.G. Douglas, Glenview, IL, Scott, Foresman and Company, 1998, pp. 7-24.

Notice:
- Titles of items published separately, such as books, magazines and pamphlets, have traditionally been underlined or italicized. Titles of items not published separately, such as poems, magazine articles and parts of books, are put in quotation marks. However, do not automatically use this pattern in citing magazine and journal articles in technical reports. Instead, see examples 7 and 8.
- Compare the publishers' names in examples 3 and 4, noting that they should be written as the publishers write them.
- When state names are needed, they are written as two capital letters.

For a book with no author's name

5. *Timber Construction Manual,* 4th ed., American Institute of Timber Construction, New York, John Wiley and Sons, Inc., 1991, p. 452.

For an encyclopedia article

6. Rihani, Ahmed, "Steam Engines and Other Heat Engines," *Encyclopedia Britannica,* 1998 ed., vol 28, pp. 903-914.

For a signed article in a professional journal

    7. Kolarov, Dimitri, Modeling the field spread of soybean mosaic virus with strain-specific monoclonal antibodies, *Agron. J.,* vol. 10, no. 3, Mar. 2000, pp. 133-138.

OR

    8. Kolarov, Dimitri, "Modeling the Field Spread of Soybean Mosaic Virus with Strain-Specific Monoclonal Antibodies," *Agronomy Journal,* vol. 10, no. 3, Mar. 2000, pp. 133-138.

Notice:

The newer, shorter form shown in example 7 omits quotation marks, caps only the first word of the article title, and abbreviates the journal title. Use this form only if you know how the journal titles for *all* the journals you are citing are abbreviated in library catalogs and indexes. If you do not know the abbreviations, use the traditional form (example 8). Whichever you choose, be consistent.

For a unsigned article in a professional journal

    9. Geisolic soil classification, *Soil Sci.,* vol. 15, no. 1, Jan. 1999, pp. 658-670.

OR

    10. "Geisolic Soil Classification," *Soil Science,* vol. 15, no. 1, Jan. 1999, pp. 658-670.

For a newspaper article with an author's name

    11. Gamboa, Glenn, "Intel Corp. Chooses `Celeron' as Tag for Computer Processors," *The Beacon Journal,* [Akron, OH] Mar. 19, 1998, p. E1.

For a newspaper article with no author's name

    12. "Businesses Protest New OSHA Guidelines," *The New York Times,* Sept. 5, 2000, sec. C, p.5.

Notice:
- If the name of the newspaper does not adequately identify where it is published, add the name of the city and/or state in brackets.
- If the newspaper uses page numbers that identify the section, a separate listing of the section letter may be omitted.

For an unsigned brochure

    13. *Computerized Coil Tracking Systems for the Steel Industry,* West Hartford, CT, ASG Systems Controls, Inc., 1999.

8.4.2 GOVERNMENT SOURCES.

For an unsigned report

    14. *Incidence of Alcohol-Related Accidents,* 1995-1999, U. S. Department of Transportation, DOT 42-89342, April 2000, pp. 62-68.

Notice:

Name the sponsoring department, agency or group, and give any identifying numbers that may help your reader locate the document.

For a commissioned report

15. "Ecological Impact of Using Prefabricated Components for Wood-Frame House Construction, The" The Timber Institute, Nashua, NH, U.S. Forset Service, FPL53, July, 1999.

Notice:

In entries beginning with a title, an article at the beginning of the title is moved to the end of the title.

For a compiled or edited report

16. DiSabato, Lorenzo, "Housing Trends in Europe: Report from a Seminar in Barcelona," technical paper #1582, complied by Harriet M. Morgan, National Residential Council, Ottawa, Canada, Oct., 2000.

8.4.3 ELECTRONIC SOURCES.
For a web site

17. "TOEFL News," *The TOEFL Center,* website, http://www.teoflcenter.com, Apr. 1, 1999.

For a videotape

18. *Radio Frequency Transmitters: Harnessing Sound for Business,* videotape, Atlanta, Vox Industries, Inc. [1999].

Notice:

When a source is undated, but the date can be obtained (if, for example, a company put no copyright date on a videotape, but can tell you when the tape was made), put the date in brackets. If no date can be determined, write [n.d.] in the date position. Avoid undated sources.

8.4.4 UNPUBLISHED SOURCES.
For a letter

19. Reno, Janet, Letter to the author, Sept. 9, 1998.

Notice:

If the letter was written to a third person, that person's name would be placed where "to the author" appears in the example above.

For an interview

20. Garcia-Bellido, Ramon, interviewed by Edyth C. Hakes, Inventure Place, Nov. 23, 1999.

For a lecture or speech

21. Burke, James, "How Renaissance Gardens Made the Carburetor Possible and other Journeys through Knowledge," University Series Lecture, Wittenberg University, Feb. 7, 2000.

OR

22. Diamant, Louwanna, "Telecommunication Satellites," class lecture, The University of Akron, May 5, 2001.

The writer who meets a challenge not addressed in the examples above should extrapolate a similar solution and should consult the instructor or editor.

You may hear the word "quote" used as a noun: "I put four quotes on the first page." However, you should know that 85% of the Usage Panel for the *American Heritage Dictionary* frowns on this usage, considering the use of "quote" as a noun (instead of "quotation") unacceptable in writing.

APPENDIX
Sample Pages From Student Reports

These sample pages are extrapolated from typical students' work and instructors' responses. Students on whose work these are based were using the *Guide* or its predecessor, *A Stylesheet for Technical Reports.* These examples will give you alternative language models for some of the key elements of a technical report and show more ways difficult details can be handled in technical reports. However, please remember that even though these are based on excellent student examples, they are not perfect. In fact, their weaknesses as well as their strengths can be instructive. Be sure to read not only the student's work but also the margin notes and instructor's responses so that you do not copy a weakness or error.

The last four pages come from two lab reports written by student Betsey Wade. They show an alternative format and two more ways to present tables. Just as important, they demonstrate an excellent resolution to one of the most confounding problems of technical writing: how to balance objectivity with personal observation. Betsey's writing is clear, straightforward and brief, presenting details of what actually happened in the lab along with an analysis of the processes and their implications.

70 *Sample Pages From Student Reports*

This is a modified block letter with indented return address, date, complementary close, signature and typed name. Note that paragraphs are not indented. It is an acceptable, conservative choice. However, this example is not spaced properly. Read the comment below.

When Brian sends his report to his employers, he will write a new letter addressed to them. A report should always be accompanied by transmittal correspondence explaining what the report is and why it is being sent. That letter can keep a writer's hard work from languishing forgotten on a shelf.

Brian should have added an enclosure notice.

---

151 Wheeler Street
Akron, OH 44311

April 29, 1998

Anna Maria Barnum
Professor, Community and Technical College
The University of Akron
Akron, OH 44325-6105

Dear Professor Barnum:

Enclosed is the final draft of my formal technical report, entitled <u>Moving Freight and Paperwork,</u> a topic you approved at the beginning of this semester. The report describes the process of freight delivery from pick-up, sorting line haul all the way to city delivery. Also covered is the parallel movement of the paperwork that accompanies the freight, from bill entry to invoicing. The report is meant to serve as an informational guide for new employees in the industry, to help them see the big picture.

I would like to thank the employees at Windfall Freight Systems for their advice and concern on this project as I completed my internship there. Many thanks also go to Professor Arthur George, my faculty advisor, who was good enough to discuss the report's organization with me. If you approve the final report, I will then send it on to my employers at Windfall for their approval.

Sincerely yours,

Brian Carnes

Brian Carnes
Student in Section 010

Figure 32 *Moving Freight,* Letter of Transmittal

Brian had a problem with spacing, both in the heading and the text. He could have solved the problem in the heading by moving the word *Professor* to the line above, so the inside address began *Professor Anna Maria Barnum* or *Anna Maria Barnum, Professor.* In the text, the solution is simple. Letters are generally single-spaced (with double-spacing between paragraphs). With single-spacing, this letter would fit neatly on the page with an appropriately generous lower margin, not a cramped one. Double-space the body of a letter only when it is very short, and always indent paragraphs when double-spacing.

CONTENTS

LIST OF FIGURES iii

ABSTRACT iv

INTRODUCTION v

DISCUSSION 1

    I. THE DIGITIZATION PROCESS 2
        Theory of Operation 2
        Versatility 3

    II. PRACTICAL ASPECTS 5
        Speed 5
        Resolution 6
        Environmental Impact 8
        Disadvantages 9
        Electronmagnetic Fields 10
        Chemical Waste 11

    III. COMMON TOOLS USED 14
        Cameras 14
        Scanners 16
        Computers 18

    IV. IMAGE MANIPULATION 20

    V. ETHICAL CONSIDERATIONS 23

CONCLUSION 25

REFERENCES 26

Convention allows this page to be titled "Contents" or "Table of Contents." For brevity and because the page is obviously a table, we recommend the shorter title.

This traditional TOC drops the outlining letters and numbers on third and fourth level headings. Instead, readers are expected to search the TOC for key words. This form is best used only for reports that are not very long or very technical.

Notice that neither the introduction nor the conclusion is numbered.

Roman numerals are traditionally set flush right, but proportional letter spacing may make this almost impossible without awkward jogging from line to line (notice IV and V). Flush left is also acceptable.

Figure 33 *Digital Photography,* Table of Contents

To create this page, coordinate the outline made earlier with the headings actually used in the report. The Table of Contents should list headings exactly as they appear in the report. In short reports the table of contents is the reader's only guide to the report's numbering system and parts. Readers locate information by consulting the table of contents instead of an index at the end.

Sample Pages From Student Reports

Technical information is complicated enough; keep your language simple and precise.

Avoid careless capitalization.

Your instructor may not catch misspelled jargon and neither will spell check. Informed readers, however, will.

Use the active voice wherever you can. This sentence is much stronger with an active voice and a more precise word choice.

Favor brevity.

ABSTRACT

Freight movement in the U.S. entails ~~a complex mixture of~~ *both* hard copy documentation and computerized data. ~~In the~~ *During* pick-up, ~~part of the process~~ a city driver collects the freight from the shipper~~,~~ *and* deliver~~ings~~ the freight and the *with a* signed ~~B~~*b*ill of ~~L~~*l*ading to the terminal. The city driver turns over the ~~B~~*b*ill of ~~L~~*l*ading to a city dispatcher, who then routes it to an ASD ~~P~~*p*rofessional within the terminal to be entered into a database. Meanwhile, dock workers sort the freight and put it into the proper ~~line hall~~ *long haul* trailers. Each ~~line hall~~ *long haul* trailer is then connected to an over-the-road tractor and pulled to its destination city, where it is unloaded, re-sorted, put into a city delivery trailer and pulled to an ultimate destination, the customer. While ~~the~~ freight ~~is moved~~ *moves* along the roads, ~~the billing process~~ *documentation* ~~is moved~~ *moves* along the wires. After the original entry into the database, the system searches for an exact match between shipper and consignee. Should this search fail ~~to find a match~~, a ~~R~~*r*esolution ~~E~~*e*xaminer manually searches the ~~B~~*b*ill. After a match has been made, a rate clerk ~~applies~~ *records* the correct ~~pricing~~ *price* ~~before sending~~ *and sends* the bill to the ~~I~~*i*nvoicing ~~D~~*d*epartment, where an invoice is printed ~~up~~ and mailed ~~to the payer of the bill~~.

iv

Figure 34 *Moving Freight,* Abstract, Edited

The amount of editing on this sample may suggest that it was poorly written. Actually, it successfully solved the most important problem of abstracting: it reviews all major parts of the report without going into much detail. The student solved that problem by writing out the process step by step in the order in which it appeared in the report. In addition, the student has made good use of the active voice, which makes language livelier and stronger. The corrected sentence (see margin note 4) demonstrates how the active voice enlivens a sentence.

DISCUSSION

I. THE DIGITIZATION PHENOMENON

~~Alternate~~ *Alternative* number systems have been recognized ~~in the world of mathematics~~ *by mathematicians* for a long time, yet most people are so accustomed to thinking and living in the base ten numbering system that at first it is difficult to ~~identify the nature of~~ *understand the base two or "binary"* ~~digital~~ numbering system used in ~~modern~~ *digital* technology. Because of the ~~very~~ nature of electricity, in which a current is either running or not running, on or off, open or closed, a binary system is useful. A binary number system consists of only two digits: zero and one, representing the two possible states.

Binary numbers are made out of these two digits, and so the numbers are written as combinations of ones and zeros. With so few variations available, the written numbers become too long to be useful to people in their arithmetical computations. For computer use, however, the numbers are ideal. Early computers used punched cards to input data: light or no light. The advent of magnetic tape and later disk storage could also use the binary system: magnetized or not magnetized.

Theory of Operation.

Digital photography is created by dividing ~~up~~ an image into

Check this page against the table of contents for this report. The headings and pagination must match.

Edit beginning paragraphs closely. You are likely to write them first, when you are not "warmed up," yet they are especially important to getting your reader off to a smooth start and a clear understanding. Here, an important improvement is in pairing "people" with "mathematicians" and "base ten" with "base two."

Notice the appropriate use of colons.

FIGURE 35 *Digital Photography*, Discussion, Page 2, Edited

Figure 35 shows the body of a report set up from a traditional outline. The working outline used Roman numerals, letters of the alphabet and Arabic numerals, but the report retains only the capital Roman numerals.

Sandra's headings should be more precise: "The Digital Phenomenon" or, better yet, "Binary Numbers and the Digital Revolution." Avoid general headings like "Theory of Operation" unless necessary; here "Digital Imaging" or "How Digital Images are Formed" would be better.

74 *Sample Pages From Student Reports*

Here is an opportunity to use active instead of passive voice : "In a color system, each pixel needs three binary numbers, one for each primary color that the computer uses in ...."

Ending a paragraph with a citation is not a good practice. It flags the reader to the fact that you have not put a conclusion on the paragraph and suggests that your control of the material may be tenuous.

Proof by eye to catch spelling, grammar and usage problems that your computer will not.

a grid of picture elements, which are called pixels. (1:201) The photography system assigns a binary number to each pixel, indicating its location and brightness. This assignment would be tedious and clumsy if done by hand, but a computer can easily deal with the very long binary numbers that result.

In a color system, three binary numbers are needed for each pixel, one number for each of the primary colors that the computer will use in varying amounts to create the proper color. Compare these colors of light: red, blue and yellow, to the four pigments of ink used in traditional color printing: yellow, magenta, cyan and black. (2:156) In the digital process, the computer converts calculated values at the different pixels back into an image by using a series of mathematical formulas.(1:233)

Versatility.

An image stored in digital form has many advantages over traditional photographs (also referred to as analogue images). Because a digital image is stored as a series of numbers, it can be copied indefinitely with no loss of quality ~~form~~ *from* the original. Unlike analogue images, which exist only as hard objects (negatives, transparencies or prints) and are therefore susceptible to environmental contamination such as dust and to eventual decomposition, digital images live forever, so long a the carrier media, such as disks, are not

3

FIGURE 36 Digital Photography, Discussion, Page 3

Wherever a fact or opinion that is not the direct observation of the writer is reported, a citation shows the source and page where the writer found the information. When proofreading a report using the numbered reference system, look for an increase in the first (or source) number in the citations. For example, only after the number 2 has appeared at least once should the number 3 appear. At any time, of course, a source number already appearing may appear again.

Sample Pages From Student Reports **75**

FINDING AN EMPIRICAL FORMULA

1 PURPOSE.

To determine the empirical formula of a hydrated binary salt by measuring the mass percents of its chemical components.

2 PROCEDURE.

The hydrated binary salt was heated in a crucible allowing the water to evaporate so the mass percent of water could be calculated. Next, a new sample of the salt was titrated to determine the mass percent of chloride in the salt. Finally, barium sulfate was collected from a sample containing both barium chloride and sulfuric acid to determine the quantity of barium ion.

3 RESULTS.

Table 1 Quantitative Water Analysis

measurement	data observed
mass crucible	10.1011g
mass salt (before heating)	1.003g
mass of salt and crucible after evaporation	10.9521g
mass dehydrated salt	0.851g
percent water	14.93%

Table 2 Quantitative Analysis of Chloride

	Trial 1	Trial 2	Trial 3	average
mass salt dissolved	0.1022g	0.1006g	0.0999g	
volume used to react completely	16.78ml	16.38ml	16.43ml	
mass Cl	0.0294g	0.0287g	0.0288g	
mass % Cl	28.8%	28.5%	28.8%	28.7%

Table 3 Quantitative Analysis of Barium

	sample A	sample B	average
mass paper and salt before filtration	1.1348g	1.1409g	
mass paper and salt after filtration	1.1051g	1.1169g	
mass barium sulfate	0.1734g	0.1777g	
mass Ba in barium chloride	0.1021g	0.1045g	
% Ba	50.3%	52.0%	51.15%

4 CALCULATIONS.

1. 10.952g (crucible + salt)

page 2

FIGURE 37 *Finding an Empirical Formula,* Discussion, Page 2

Betsey's chemistry professor required using "Discussion" as the report's fourth heading, so the report's title was used in place of the word "Discussion."

Notice the fragment below the first subheading. Some instructors will accept this, but most authorities recommend repeating the subhead or a similar phrase: "The purpose of this lab experiment was to determine...."

Use "Analysis of Water" to make this heading parallel with the others.

Notice the Arabic numeral and flush left heading for "Table 1...." Betsey favored lower case for the headings, which is good, although "Trial 1," ect. should also have been lower case.

The report includes 25 formulas, so it is appropriate to number them.

These three tables show an appropriate use of gray shading to set off each table and separate the column headings.

76 *Sample Pages From Student Reports*

These 2-part numbers label subsets of formula six; they are not part of the outline.

Ending each discussion of possible error by summarizing what it would have done to the results is an example of good repetition. It shows, in a structured way, that Betsey understands the implications of the experiment.

6.4 $\dfrac{0.829 \text{ mol } H_2O}{0.3724 \text{ mol}} = 2.26 \text{ mol}$

6.5 $x = 1, y = 2, z = 2$ Ba Cl H_2O

5 DISCUSSION.

5.1 Errors in Table 1.

1. An execution error could have been in not allowing the water in the salt to evaporate. If there were still water in my sample, the mass of water would be too high and therefore would cause the mass percent of water to be lower than it actually is. So, the mole value for z should have actually been higher. (The error would cause z to be too low).

2. An execution error could have occurred if I accidentally allowed some of my salt to get on the tongs. I used the tongs to pick up the crucible, and if, by chance, the tongs slipped and picked up some of the dehydrated salt, my mass of salt would have been too low, because some of my salt would have been missing. This would cause the mass percent of water to be too high and the mole value for z would have actually been lower than I calculated. (The error would cause z to be too high.)

3. A methodological error could have occurred when we allowed our crucible to cool. We heated it to allow all the water in the salt to evaporate. However, during the 15 minutes that the sample was set aside to cool, new water may have precipitated on the salt. This would cause our mass of salt without water to be too high, and would therefore cause the mass percent of water to be lower than it actually is. In this case the mole value for z should actually have been higher. (The error would make z too low.)

5.2 Errors in Table 2.

1. An execution error could have been in over-titrating. I may have over-titrated because it was so hard to tell the color change in the solution. If I over-titrated, my volume would have been too high, which would cause my

page 4

FIGURE 38 *Finding an Empirical Formula,* Discussion, Page 4

This lab report is exceptionally clear and direct. When Betsey means herself, she says "I." When she means herself and her lab partner, she uses "we." When referring to a process or a piece of equipment, she writes in the third person. This tells the lab instructor what actually happened in the lab.

SYNTHESIS AND ANALYSIS OF COORDINATION COMPOUNDS

1 PURPOSE.

The purpose was to synthesize both a copper and a nickel complex that each has ammonia ligands and then to precipitate the salt out of the solution to find an empirical formula for the nickel and copper complexes.

2 PROCEDURE.

The directions in the lab manual for *Synthesis and Analysis of Coordination Compounds* on pages 12-15 were followed. We filtered the coordination compound using vacuum filtration, washed the crystals, and then titrated them. The titrations for the experiment were performed using the MPLI software with pH probe. From the titration, percent composition of NH3 in each sample was found and the empirical formulas were determined from the percent composition. The percent yield was then calculated.

3 DATA AND RESULTS.

Table 1 Standardization Data, MPLI Endpoint Detection

Trial 1, volume, mL	8.66
Trial 2, volume, mL	20.02
Trial 3, volume, mL	20.11
Average volume, mL	19.60
Standard deviation	0.8124
Percent relative standard deviation	4.14%
Molarity of NaOH, M*	0.25
Moles of NaOH titrated	0.00625
Moles of HCl titrated	0.00625
Molarity of HCl	0.319

Betsey used an Excel table (which automates calculations) and imported it into her Word document. If you have such programs available, but they impose certain formatting, discuss any concerns you may have with your instructor.

page 2

FIGURE 39 *Synthesis and Analysis,* Discussion, Page 2

78 *Sample Pages From Student Reports*

CONCLUSION

Our percent yield was greater than 100%, and at first this seemed impossible. However, because we rounded up our number of moles of NH3 in the copper complex from 3.8 to 4 and in the nickel complex from 5.8 to 6, we calculated more grams in our theoretical yield than we actually found, which could make up for this discrepancy. Also, we may not have weighed out exactly 25 grams of each of our coordination compounds, or we could have lost some of our compound in the transfer process.

I found this lab to be a very interesting introduction to ligands in the forming of coordination compounds and an interesting way to learn a new method of measuring pH and using pH titrations to calculate an empirical formula. It was fascinating to see how by knowing only the molarity of the NaOH, we could standardize our HCl and using the mass of our coordination compounds and the volume of HCl to titrate them, we were able to determine the empirical formula of our complexes.

page 4

FIGURE 40 *Synthesis and Analysis,* Conclusion

In the first paragraph, Betsey speaks for herself and her lab partner, tracing their thought processes as they made sense of their findings. The second paragraph begins with Betsey speaking for herself, so the pronoun changes. The paragraphing is also appropriate as it separates the analysis of findings from the writer's more personal reaction to the overall importance of the lab work.